GSM
Switching, Services and Protocols

GSM

Switching, Services and Protocols

Jörg Eberspächer
and
Hans-Jörg Vögel
Technische Universität München, Germany

JOHN WILEY & SONS
Chichester · New York · Weinheim · Brisbane · Singapore · Toronto

Other Wiley Editorial Offices

John Wiley & Sons, Inc., 605 Third Avenue,
New York, NY 10158-0012, USA

WILEY-VCH GmbH, Pappelallee 3,
D-69469 Weinheim, Germany

Jacaranda Wiley Ltd, 33 Park Road, Milton,
Queensland 4064, Australia

John Wiley & Sons (Asia) Pte Ltd, 2 Clementi Loop #02-01,
Jin Xing Distripark, Singapore 129809

John Wiley & Sons (Canada) Ltd, 22 Worcester Road,
Rexdale, Ontario M9W 1L1, Canada

Library of Congress Cataloging-in-Publication Data

Eberspächer, J. (Jörg)
    [GSM, Global System for Mobile Communication. English]
    GSM switching, services, and protocol / Jörg Eberspächer & Hans
  –Jörg Vögel.
      p.   cm.
    Includes bibliographical references and index.
    ISBN 0-471-98278-4 (alk. paper)
    1. Global system for mobile communications.   I. Vogel, Hans-Jörg.
  II. Title.
  TK5103.483.E2413   1998
  621.382 — dc21   98-30914
                                      CIP

British Library Cataloguing in Publication Data

A catalogue record for this book is available from the British Library

ISBN 0-471-98278-4

Produced from PostScript files supplied by the author.
Printed and bound in Great Britain by Biddles Ltd, Guildford, UK.
This book is printed on acid-free paper responsibly manufactured from sustainable forestry, in which at least two trees are planted for each one used for paper production.

Contents

Preface

GSM is much more than the acronym of Global System for Mobile Communication; it stands for an extraordinarily successful stage of development in modern information technology. GSM means a new dimension for more than 50 million users − and there are more and more every day − a dimension of personal communication. Today GSM is deployed in more than 100 countries and by over 220 network operators, many of them outside Europe. The mobile telephone has advanced from status symbol to useful appliance, not only in business but also in private everyday life. Its principal use is for wireless telephony, but GSM data communication is increasingly gaining importance.

This modern digital system for mobile communication is based on a set of standards, which were worked out in Europe and can now be considered truly global. Many of the new standardization initiatives of GSM Phase 2+ are in fact coming from outside of Europe. Depending on locally available frequency bands, different GSM air interfaces are defined (e.g. for 900 MHz, 1800 MHz, and 1900 MHz). However, architecture and protocols, in particular for user–network signaling and global roaming are identical in all networks. Thus, GSM enables worldwide development, manufacturing and marketing of innovative products, that stand up well under competition.

GSM also stands for complexity. Whether in the terminals or the exchange equipment, whether in hardware or software, GSM technology is extraordinarily involved and extensive; certainly the most complex communication systems by themselves comprise the standards published by the European Telecommunication Standards Institute (ETSI).

This book arose from an effort to explain and illustrate the essential technical principles of GSM in spite of this complexity, and to show the interrelations between the different subfunctions in a better way than is possible in the framework of standards. Points of crystallization were provided by our course "Communication Networks 2" at the Munich University of Technology as well as our GSM lab course, which requires the students to prepare by studying an extensive GSM manuscript. This lab course is also part of the English graduate program in "communications engineering" at our university which is leading to an MSc degree. The foundation of this book is, however, in the ETSI standards themselves (besides some scientific publications), which were, on one hand, "boiled down" in this book and, on the other hand, augmented by explanations and interpretations.

The book is intended for all those who want to acquire a deeper knowledge of the complex GSM system without losing their way in the detail and wording of the standards. Addressed are the students of electrical engineering, computer science, and information technology at universities and technical institutes, those in industry or network operations who use and apply the technology, but also researchers who want to gain insight into the architecture and functional operation of the GSM system.

In accordance with the publisher and editors, our book presents the entire architecture of GSM with concentration on the communication protocols, the exchange technology, and the realization of services. The most important principles of the GSM transmission technology are also included in order to give a rounded treatment. Those who are involved with the implementation of GSM systems should find the book to be a useful start and they should find adequate guidance on the standards. The study of the standards is also recommended when there are doubts about the latest issues of the ETSI standards, for with this book we had to consider the standards to be "frozen" in their state as of summer 1997.

The authors especially thank Professor Martin Bossert (Ulm University) for many helpful hints and clarifying discussions. We are very grateful to Professor Gottfried R. Luderer (Arizona State University, Tempe, AZ) for the translation of the German version of the book as well as for the critical technical review of the manuscript and numerous proposals for improvement. It was his strong commitment and determined translation work, which made this book possible. We also give our cordial thanks to the people at Wiley for initiating this book and for the smooth cooperation. Their support in every phase of the project was critical to its speedy production and publication.

The authors are grateful in advance for any kind of response to this book. Readers should address us (wireless or over guided media), preferably via email.

Munich, July 1998   Jörg Eberspächer
                                                      Joerg.Eberspaecher@ei.tum.de

                                                      Hans-Jörg Vögel
                                                      Hans-Joerg.Voegel@ei.tum.de

1 Introduction

1.1 Digital, Mobile, Global: Evolution of Networks

Communication everywhere, with everybody, and at any time – we have come much closer to this goal during the last few years. Digitalization of communication systems, enormous progress in microelectronics, computers, and software technology, inventions of efficient algorithms and procedures for compression, security, and processing of all kinds of signals, as well as the development of flexible communication protocols have been important prerequisites for this progress. Today technology is available which enables the realization of high-performance and cost-effective communication systems for many application areas. In the field of fixed networks – where the end systems (user equipment) are connected to the network over a line (copper two-wire line, coaxial cable, glass fiber) – the Integrated Services Digital Network (ISDN) system is steadily gaining importance, especially in Europe. Coming from the USA, the packet-switched Internet has made a triumphant advance, and cable and satellite networks are also used increasingly for individual communication.

The largest technological and organizational challenge is, however, the support of subscriber mobility. In this case, the distinction has to be made between two kinds of mobility:

- terminal mobility
- personal mobility

In the case of *terminal mobility*, the subscriber is connected to the network in a wireless way – via radio or light waves – and can move with his or her terminal relatively freely, even during a communication connection. The degree of mobility depends on the kind of mobile radio network. In the case of a cordless telephone for the home, the requirements are much less critical than for the mobile telephone which can be used on the road, in a car or train. If mobility is to be supported across the whole network (or country) or even beyond the network or national boundaries, additional switching technology and administrative functions are required, to let subscribers communicate in wireless mode outside of their "home areas".

Such extended network functions are also needed to realize *personal mobility* and universal reachability. This is understood to comprise the possibility of location-independent use of all kinds of telecommunication services – including and especially in fixed networks. The user identifies himself or herself (the person), e.g. with the help of a chipcard, at the place where he or she is currently staying and has access to the network. There, the same communication services can be used as at home, limited only by the properties of the local network or terminal used. A worldwide unique and uniform addressing is an important requirement.

In the digital mobile communication system GSM *(Global System for Mobile Communication)*, which is the subject of this book, terminal mobility is the predominant issue. Wireless communication has become possible with GSM in any town, any country, and even – already partially realized – on any continent.

GSM technology, however, contains already the essential "intelligent" functions for the support of personal mobility, especially with regard to user identification and authentication, and for the localization and administration of mobile users. Here it is often overlooked that in mobile communication by far the largest part of the communication occurs over the fixed network part, which is needed to interconnect the radio stations (base stations).

Therefore it is no surprise that in the course of further development and evolution of the telecommunication networks, a lot of thought is given to the convergence of fixed and mobile networks; the worldwide standardization of the "universal" network of the future has already progressed quite far. Europe has good chances once more to play a leading role with her concept of the *Universal Mobile Telecommunication System* (UMTS).

1.2   Classification of Mobile Communication Systems

This book deals almost exclusively with GSM; however, GSM is only one of many facets of modern mobile communication. Figure 1.1 shows the whole spectrum of today's and – as far as can be seen – future mobile communication systems. Unidirectional message systems (paging) have proliferated; they achieve very cost-effective reachability with wide-area coverage.

For the bidirectional – and hence genuine – communication systems, the simplest variant is the cordless telephone (in Europe especially the DECT standard) with very limited mobility. This technology is also employed for the expansion of digital PBXs with mobile extensions. A related concept is *Radio in the Local Loop* (RTLL) or *Wireless Local Loop* (WLL).

This concept requires only limited mobility. Wireless telephone booths have received less public acceptance. These are systems with restricted communication capability (only outgoing calls are possible). Cellular systems, however, have been extremely successful. First, they are offered in the form of special (trunked) mobile radio systems – in digital form with the European standard *Trans European Trunked Radio* (TETRA) – which are used for business applications like fleet control. On the other hand, cellular systems are used predominantly for public mass communication. They had an early success with analog systems like the *Advance Mobile Phone System* (AMPS) in America, the *Nordic Mobile Telephone* (NMT) in Scandinavia, or the *C-Netz* in Germany. Founded on the digital system GSM with its variants for 900 MHz, 1800 MHz, and 1900 MHz, which was developed in Europe, a market with millions of subscribers worldwide was generated, and it represents an important economic force. A strongly contributing factor to this rapid development of markets and technologies has been the deregulation of the telecommunication markets, i.e. it allowed the establishment of new network operators.

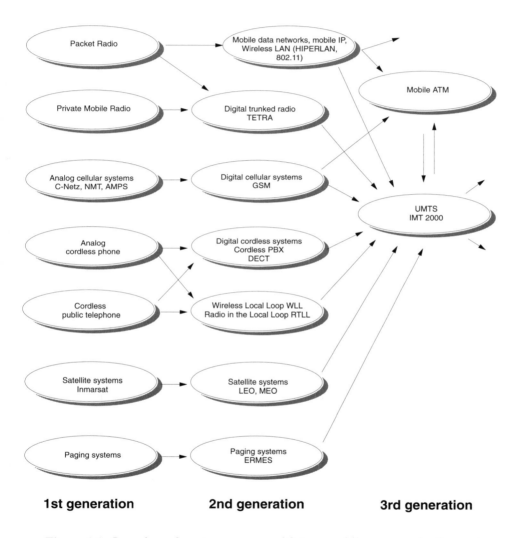

Figure 1.1: Overview of contemporary and future mobile communication systems

Another competing or supplementing technology is satellite communication, which also promises to offer global, and in the long term even broadband, communication services. Future systems are based on *Low Earth Orbiting* (LEO) or *Medium Earth Orbiting* (MEO) satellites.

Local area networks (LANs), which are in widespread use within buildings, have also been augmented with mobility functions: wireless LANs are offered and are being standardized by ETSI HIPERLAN and by IEEE 802.11. The efforts of Internet "mobilization" Mobile

IP are also worth mentioning in this context. Mobile data and multimedia communication are going to see a further strong innovation impulse with the development of wireless Mobile ATM systems based on the exchange technology *Asynchronous Transfer Mode* (ATM).

The path to the future universal telecommunication networks (UMTS, IMT2000) has been opened with the realization, mentioned above, of the personal communication services, *Universal Personal Telecommunication* (UPT), based on intelligent networks. Which technology will ultimately prevail and whether the Holy Grail of a uniform worldwide universal network will ever be achieved, is still an open question today.

In any case, GSM will remain for many years the technological base for mobile communication, and it continues to open up new application areas. At the moment, the area of intelligent vehicular applications seems to be particularly attractive, where GSM data communication is serving well for transmission of traffic-related information. The techniques and procedures presented in this book are the foundation for such innovative applications.

1.3   Some GSM History

In 1982 the development was started of a pan-European standard for digital cellular mobile radio by the *Groupe Spécial Mobile* of the CEPT (Conférence Européenne des Administrations des Postes et des Télécommunications). Initially, the acronym GSM was derived from the name of this group. After the founding of the European standardization institute ETSI (*European Telecommunication Standards Institute*), the GSM group became a Technical Committee of ETSI in 1989. After the rapid worldwide proliferation of GSM networks, the name has been reinterpreted as *Global System for Mobile Communication*. Today this is the official designation.

After a series of incompatible analog networks had been introduced in parallel in Europe – e.g. *Total Access Communication System* (TACS) in the UK, *Nordic Mobile Telephone* (NMT) in Scandinavia, and the *C-Netz* in Germany – work on the definition of a Europewide standard for digital mobile radio was started in the late 1980s. The GSM was founded, which developed a set of *Technical Recommendations* and presented them to ETSI for approval. These proposals were produced by the *Special Mobile Group* (SMG) in working groups called *Sub Technical Committees* (STCs), with the following division of tasks: service aspects in STC SMG 1, radio aspects in STC SMG2, network aspects in STC SMG 3, data services in STC SMG 4, and network management in STC SMG 6. STC SMG 5 deals with future networks and is responsible for the standardization of the next generation of the European mobile radio system, the *Universal Mobile Telecommunication System* (UMTS). These proposals were published as GSM Recommendations and comprise about 130 single documents of more than 5000 pages in total.

After the official start of the GSM networks during the summer of 1992 (Table 1.1), the number of subscribers has increased rapidly, such that during the fall of 1993 already far more than one million subscribers made calls in GSM networks, more than 80% of these in Germany alone. On a global scale, the GSM standard also received very fast recognition, as evident from the fact that at the end of 1993 several commercial GSM networks started operation outside of Europe, in Australia, Hong Kong, and New Zealand. GSM has also been introduced in Brunei, Cameroon, Iran, South Africa, Syria, Thailand, the USA and the United Arab Emirates. Whereas the majority of the GSM networks operate in the 900 MHz band (GSM900), there are also networks operating in the 1800 MHz band (GSM1800) – *Personal Communication Network* (PCN), *Digital Communication System* (DCS1800) – and in the United States in the 1900 MHz band (GSM1900) – *Personal Communication System* (PCS). These networks use almost completely identical technology and architecture; they differ essentially only in the radio frequencies used and the pertinent high-frequency technology, such that synergy effects can be taken advantage of, and the mobile exchanges can be constructed with standard components.

Table 1.1: Time history — milestones in the evolution of GSM

Year	Event
1982	Groupe Spécial Mobile established by the CEPT.
1987	Essential elements of wireless transmission are specified, based on prototype evaluation (1986). Memorandum of Understanding (MoU) Association founded in September with 13 members from 12 countries.
1989	GSM becomes an ETSI Technical Committee (TC).
1990	The Phase 1 GSM900 specifications (designed 1987 through 1990) are frozen. Adaption to DCS1800 commences.
1991	First GSM networks launched. The DCS1800 specifications are finalized.
1992	Most European GSM networks turn commercial by offering voice communication services. Some 13 networks in 7 countries are "on air" by the end of the year.
1993	First roaming agreements in effect By the end of 1993, 32 networks in 18 countries are operational.
1994	Data transmission capabilities launched. The number of networks rises to 69 in 43 different countries by the end of 1994.
1995	MoU counts 156 members from 86 countries. After the GSM standardization Phase 2 including adaptations and modifications for the PCS1900 (Personal Communication System) is passed, the first PCS1900 Network is launched in the USA. Facsimile, data and SMS roaming starts. Videosignals are transmitted via GSM for demonstration purposes. An estimated 50 000 GSM base stations are in use all over the world.
1996	January: 120 networks in 71 countries operational. June: 133 networks in 81 countries operational.
1997	July: 200 GSM networks from 109 countries operational, amounting to 44 million subscribers worldwide.

In parallel to the standardization efforts of ETSI, already in 1987 the then existing prospective GSM network operators and the national administrations joined in a group whose members signed a common *Memorandum of Understanding* (MoU). The *MoU Association* was supposed to form a base for allowing the transnational operation of mobile stations using internationally standardized interfaces. In the middle of 1997, ten years after its founding, the GSM MoU had 239 members which operated GSM networks in 109 countries. In total, there were (at the time of writing) 200 GSM networks in operation with about 44 million subscribers in all three frequency bands (900 MHz, 1800 MHz, 1900 MHz). The proportion of GSM of the worldwide radio communication market is thus about 28%, and it is growing by about 2 million new GSM customers each month. (Information source: web page of GSM MoU, http://www.gsmworld.com).

All of these networks have implemented Phase 1 of the GSM standard, or the later defined PCN / PCS version of it. In many places, additional services and service characteristics of GSM Phase 2 have also been realized. However, Phase 1 is essentially the basis for the following presentation.

2 The Mobile Radio Channel and the Cellular Principle

Many measures, functions and protocols in digital mobile radio networks are based on the properties of the radio channel and its specific qualities in contrast to information transmission through guided media. For the understanding of digital mobile radio networks it is therefore absolutely necessary to know a few related basic principles. For this reason, the most important fundamentals of the radio channel and of cellular and transmission technology will be presented and briefly explained in the following. For a more detailed treatment, see the extensive literature [25][26][27][30].

2.1   Characteristics of the Mobile Radio Channel

The electromagnetic wave of the radio signal propagates under ideal conditions in free space in a radial-symmetric pattern, i.e. the received power P_{Ef} decreases with the square of the distance L from the transmitter:

$$P_{Ef} \sim \frac{1}{L^2}$$

These idealized conditions do not apply in terrestrial mobile radio. The signal is scattered and reflected, for example, at natural obstacles like mountains, vegetation, or water surfaces. The direct and reflected signal components are then superimposed at the receiver.

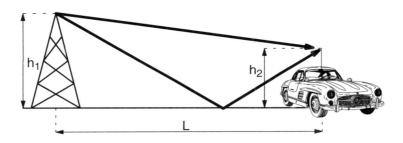

Figure 2.1: Simplified two-path model of radio propagation

This multipath propagation can already be explained quite well with a simple two-path model (Figure 2.1). With this model, one can show that the received power decreases much more than with the square of the distance from the transmitter.

We can approximate the received power by considering the direct path and only one reflected path (two-path propagation) [25]:

$$P_E = P_0 \frac{4}{(4\pi L/\lambda)^2} \left(\frac{2\pi h_1 h_2}{\lambda L}\right)^2 = P_0 \left(\frac{h_1 h_2}{L^2}\right)^2$$

and we obtain, under the simplified assumptions of the two-path propagation model, from Figure 2.1 a propagation loss of 40 dB per decade:

$$a_E = \frac{P_{E2}}{P_{E1}} = \left(\frac{L_1}{L_2}\right)^4 \qquad a_E = 40 \log\left(\frac{L_1}{L_2}\right) \text{ [dB]}$$

In reality, the propagation loss depends on the propagation coefficient γ, which is determined by environmental conditions:

$$P_E \sim L^{-\gamma}; \quad 2 \le \gamma \le 5$$

In addition, propagation losses are also frequency dependent, i.e. in a simplified way, propagation attenuation increases disproportionately with the frequency.

However, multipath propagation does not only incur a disproportionately high path propagation loss. The different signal components reaching the receiver have traveled different distances by virtue of dispersion, infraction, and multiple reflections, hence they show different phase shifts. On one hand, there is the advantage of multipath propagation, that a partial signal can be received even if there is no direct path, i.e. there is no line of sight between mobile and base station. On the other hand, there is a serious disadvantage: the superpositions of the individual signal components having different phase shifts with regard to the direct path can lead, in the worst cases, to cancellations, i.e. the received signal level shows severe disruptions. This phenomenon is called *fading*. In contrast to this fast fading caused by multipath propagation, there is slow fading caused by shadowing. Along the way traveled by a mobile station, multipath fading can cause significant variations of the received signal level (Figure 2.2). Periodically occurring signal breaks at a distance of about half a wavelength are typically 30 to 40 dB. The smaller the transmission bandwidth of the mobile radio system, the stronger the signal breaks — at a bandwidth of about 200 kHz per channel this effect is still very visible [28].

Furthermore, the fading dips become flatter as one of the multipath components becomes stronger and more pronounced. Such a dominant signal component arises, for example, in the case of a direct line of sight between mobile and base station, but it can also occur under other conditions. If such a dominant signal component exists, we talk of a Rice channel and Ricean fading, respectively. (S. O. Rice was an American scientist and mathematician.) Otherwise, if all multipath components suffer from approximately equal propagation conditions, we talk of Rayleigh fading. (J. W. Strutt, 3rd Baron Rayleigh, was a British physicist, Nobel prizewinner.)

During certain time periods or time slots, the transmission can be heavily impacted because of fading or can be entirely impossible, whereas other time slots may be undisturbed. The results of this effect within the user data are alternating phases, which show either a high or low bit error rate, which is leading to error bursts. The channel thus has memory in contrast to the statistically independent bit errors in memoryless symmetric binary channels.

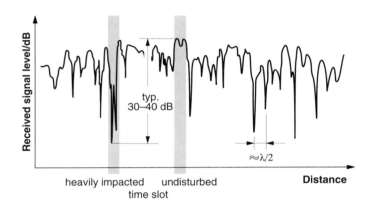

Figure 2.2: Typical signal in a channel with Rayleigh fading

The signal level observed at a specific location is also determined by the phase shift of the multipath signal components. This phase shift depends on the wavelength of the signal, and thus the signal level at a fixed location is also dependent on the transmission frequency. Therefore the fading phenomena in radio communication are also frequency specific. If the bandwidth of the mobile radio channel is small (narrowband signal), then the whole frequency band of this channel is subject to the same propagation conditions, and the mobile radio channel is considered *frequency-nonselective*. Depending on location (Figure 2.2) and the spectral range (Figure 2.3), the received signal level of the channel, however, can vary considerably. On the other hand, if the bandwidth of a channel is large (broadband signal), the individual frequencies suffer from different degrees of fading (Figure 2.3) and this is called a *frequency-selective* channel [29][32]. Signal breaks because of frequency-selective fading along a signal path are much less frequent for a broadband signal than for a narrowband signal, because the fading holes only shift within the band and the received total signal energy remains relatively constant [28].

Besides frequency-selective fading, the different propagation times of the individual multipath components also cause time dispersion on their propagation paths. Therefore, signal distortions can occur due to interference of one symbol with its neighboring symbols ("intersymbol interference"). These distortions depend first on the spread experienced by a pulse on the mobile channel, and second on the duration of the symbol or of the interval between symbols. Typical multipath channel delays have a range from half a microsecond in urban areas to about 16–20 microseconds in mountainous terrain, i.e. a transmitted pulse generates several echoes which reach the receiver with delays of up to 20 microseconds. In digital mobile radio systems with typical symbol durations of a few microseconds, this can lead to smearing of individual pulses over several symbol durations.

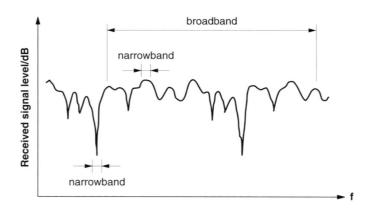

Figure 2.3: Frequency selectivity of a mobile radio channel

In contrast to wireline transmission, the mobile radio channel is a very bad transmission medium of highly variable quality. This can go so far that the channel cuts out for short periods (deep fading holes) or that single sections in the data stream are so much interfered with (bit error rate typically 10^{-2} or 10^{-1}), that unprotected transmission without further protection or correction measures is hardly possible. Therefore, mobile information transport requires additional, often very extensive measures, which compensate for the effects of multipath propagation. First, an equalizer is necessary, which attempts to eliminate the signal distortions caused by intersymbol interference. The operational principle of such an equalizer for mobile radio is based on the estimation of the channel pulse response to periodically transmitted, well-known bit patterns, known as the training sequences [27][30]. This allows the determination of the time dispersion of the channel and its compensation. The performance of the equalizer has a significant effect on the quality of the digital transmission. On the other hand, for the efficient transmission in digital mobile radio, channel coding measures are indispensable, such as forward error correction with error-correcting codes, which allows reduction of the effective bit error rate to a tolerable value (about 10^{-5} to 10^{-6}). Further important measures are control of the transmitter power and algorithms for the compensation of signal interruptions in fading, which may be of such a short duration that a disconnection of the call would not be appropriate.

2.2 Separation of Directions and Duplex Transmission

The most frequent form of communication is the bidirectional communication which allows simultaneous transmitting and receiving. A system capable of doing this is called full-duplex. One can also achieve full-duplex capability, if sending and receiving do not occur simultaneously but switching between both phases is done so fast that it is not noticed by the subscriber, i.e. both directions can be used quasi-simultaneously. Modern digital mobile radio systems are always full-duplex capable. Essentially, two basic duplex procedures are employed: *Frequency Division Duplex* (FDD) using different frequency bands in each direction, and *Time Division Duplex* (TDD) which periodically switches the direction of transmission.

The frequency duplex procedure has been used already in analog mobile radio systems and is still used in digital systems, too. For the communication between mobile and base station, the available frequency band is split into two partial bands, to enable simultaneous sending and receiving. One partial band is assigned as uplink (from mobile to base station) and the other partial band is assigned as downlink (from base to mobile station):

- transmission band of the mobile station = receiving band of the base station (uplink)
- receiving band of the mobile station = transmission band of the base station (downlink)

To achieve good separation between both directions, the partial bands must be a sufficient frequency distance apart, i.e. the frequency pairs of a connection assigned to uplink and downlink must have this distance band between them. Usually, the same antenna is used for sending and receiving. A duplexing unit is then used for the directional separation, consisting essentially of two narrowband filters with steep flanks (Figure 2.4). These filters, however, can not be integrated, so pure frequency duplexing is not appropriate for systems with small compact equipment [32].

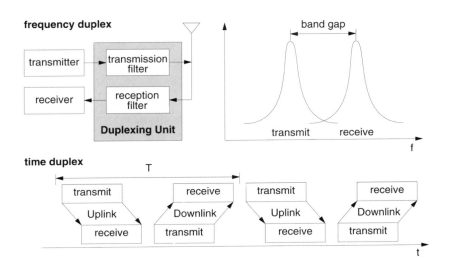

Figure 2.4: Frequency and time duplex (schematic)

Time duplexing is therefore a good alternative, especially in digital systems with time division multiple access. Transmitter and receiver operate in this case only quasi-simultaneously at different points in time; i.e. the directional separation is achieved by switching in time between transmission and reception, and thus no duplexing unit is required. Switching occurs frequently enough that the communication appears to be over a quasi-simultaneous full-duplex connection. However, out of the periodic interval T available for the transmission of a time slot only a small part can be used, so that a time duplex system requires more than twice the bit rate of a frequency duplex system.

2.3   Multiple Access Procedures

The radio channel is a communication medium shared by many subscribers in one cell. Mobile stations compete with one another for the frequency resource to transmit their information streams. Without any other measures to control simultaneous access of several users, collisions can occur (multiple access problem). Since collisions are very undesirable for a connection-oriented communication like mobile telephony, the individual subscribers / mobile stations must be assigned dedicated channels on demand. In order to divide the available physical resources of a mobile system, i.e. the frequency bands, into voice channels, special multiple access procedures are used which will be presented in the following (Figure 2.5).

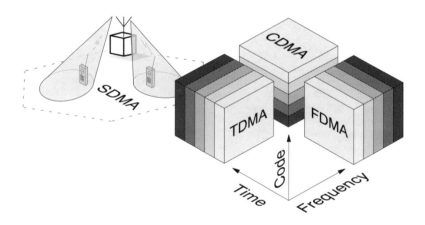

Figure 2.5: Multiple access procedures

2.3.1   Frequency Division Multiple Access (FDMA)

Frequency Division Multiple Access (FDMA) is one of the most common multiple access procedures. The frequency band is divided into channels of equal bandwidth such that each conversation is carried on a different frequency (Figure 2.6). Best suited to analog mobile radio, FDMA systems include the *C-Netz* in Germany, TACS in the UK, and AMPS in the USA. In the *C-Netz*, two frequency bands of 4.44 MHz each are subdivided into 222 individual communication channels at 20 kHz bandwidth. The effort in the base station to realize a frequency division multiple access system is very high. Even though the required hardware components are relatively simple, each channel needs its own transceiving unit. Furthermore, the tolerance requirements for the high-frequency networks and the linearity of the amplifiers in the transmitter stages of the base station are quite high, since a large number of channels need to be amplified and transmitted together [29][32]. One also needs a duplexing unit with filters for the transmitter and receiver units to enable full-duplex operation, which makes it nearly impossible to build small, compact mobile stations, since the required narrowband filters can hardly be realized with integrated circuits.

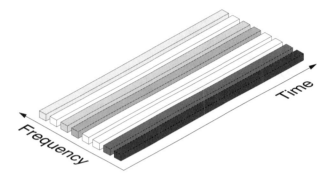

Figure 2.6: Channels of an FDMA system (schematic)

2.3.2   Time Division Multiple Access (TDMA)

Time Division Multiple Access (TDMA) is a more expensive technique, for it needs a highly accurate synchronization between transmitter and receiver. The TDMA technique is used in digital mobile radio systems. The individual mobile stations are cyclically assigned a frequency for exclusive use only for the duration of a time slot. Furthermore, in most cases one assigns not the whole system bandwidth for a time slot to one station, but subdivides the system frequency range into subbands, and then uses TDMA for multiple access to each subband. The subbands are known as carrier frequencies, and the mobile systems using this technique are designated as multicarrier systems (not to be confused with multicarrier modulation). The pan-European digital system GSM employs such a combination of FDMA and TDMA, it is a multicarrier TDMA system. A frequency range of 25 MHz holds 124 single channels (carrier frequencies) of 200 kHz bandwidth each, with each of these frequency channels containing again 8 TDMA conversation channels.

Thus the sequence of time slots assigned to a mobile station represents the physical channels of a TDMA system. In each time slot, the mobile station transmits a data burst. The period assigned to a time slot for a mobile station thus also determines the number of TDMA channels on a carrier frequency. The time slots of one period are combined into a so-called TDMA frame. Figure 2.7 shows five channels in a TDMA system with a period of four time slots and three carrier frequencies.

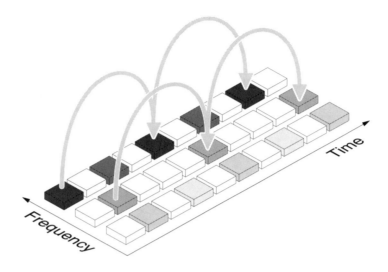

Figure 2.7: TDMA channels on multiple carrier frequencies

The TDMA signal transmitted on a carrier frequency in general requires more bandwidth than an FDMA signal, since because of multiple time use, the gross data rate has to be correspondingly higher. For example, GSM systems employ a gross data rate (modulation data rate) of 271 kbit/s on a subband of 200 kHz, which amounts to 33.9 kbit/s for each of the eight time slots.

Especially narrowband systems suffer from time- and frequency-selective fading (Figures 2.2 and 2.3) as already mentioned. In addition, there are also frequency-selective cochannel interferences, which can contribute to the deterioration of the transmission quality. In a TDMA system, this leads to the phenomenon that the channel can be very good during one time slot, and very bad during the next time slot when some bursts are strongly interfered with. On the other hand, a TDMA system offers very good opportunities to attack and drastically reduce such frequency-selective interference by introducing a frequency-hopping technique. With this technique, each burst of a TDMA channel is transmitted on a different frequency (Figure 2.8).

In this technique, selective interference on one frequency at worst hits only every ith time slot, if there are i frequencies available for hopping. Thus the signal transmitted by a frequency hopping technique uses frequency diversity. Of course, the hopping sequences must be orthogonal, i.e. one must ascertain that two stations transmitting in the same time slot don't use the same frequency. Since the duration of a hopping period is long compared to the duration of a symbol, this technique is called *slow frequency hopping*. With fast frequency hopping, the hopping period is shorter than a time slot and is of the order of a single symbol duration or even less. This technique then belongs already to the *spread spectrum techniques* of the family of code division multiple access techniques, *Frequency Hopping CDMA* (FH-CDMA); see Section 2.3.3.

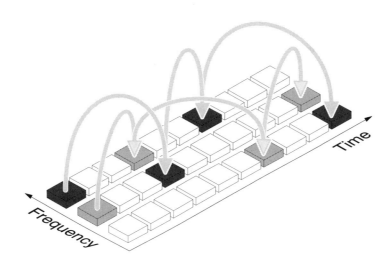

Figure 2.8: TDMA with use of frequency hopping technique

As mentioned above, for TDM access, a precise synchronization between mobile and base station is necessary. This synchronization becomes even more complex through the mobility of the subscribers, because they can stay at varying distances from the base station and their signals thus incur varying propagation times. First, the basic problem is to determine the exact moment when to transmit. This is typically achieved by using one of the signals as a time reference, like the signal from the base station (downlink, Figure 2.9). On receiving the TDMA frame from the base station, the mobile can synchronize and transmit time slot synchronously with an additional time offset (e.g. three time slots in Figure 2.9).

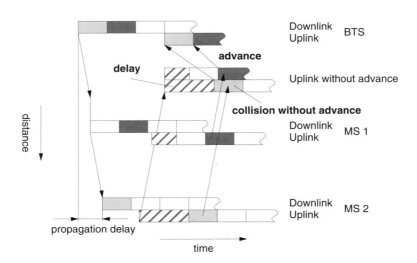

Figure 2.9: Differences in propagation delays and synchronization in TDMA systems

Another problem is the propagation time of the signals, so far ignored. It also depends on the variable distance of the mobile station from the base. These propagation times are the reason why the signals on the uplink arrive not frame-synchronized at the base, but with variable delays. If these delays are not compensated, collisions of adjacent time slots can occur (Figure 2.9). In principle, the mobile stations must therefore advance the time-offset between reception and transmission, i.e. the start of sending, so much that the signals arrive frame-synchronous at the base station.

2.3.3   Code Division Multiple Access (CDMA)

Systems with *Code Division Multiple Access* (CDMA) are broadband systems, in which each subscriber uses the whole system bandwidth (similar to TDMA) for the complete duration of the connection (similar to FDMA). Furthermore, usage is not exclusive, i.e. all the subscribers in a cell use the same frequency band simultaneously. To separate the signals, the subscribers are assigned orthogonal codes.

The basis of CDMA is a band-spreading or spread spectrum technique. The signal of one subscriber is spread spectrally over a multiple of its original bandwidth. Typically, spreading factors are between 10 and 1000; they generate a broadband signal for transmission from the narrowband signal, and this is less sensitive to frequency-selective interference and disturbances. Furthermore, the spectral power density is decreased by band spreading, and communication is even possible below the noise threshold [32].

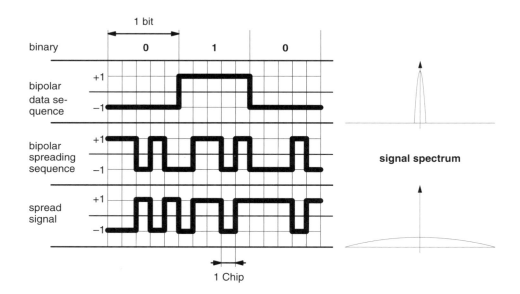

Figure 2.10: Principle of spread spectrum technique for *direct sequence CDMA*

A common spread-spectrum procedure is the *direct sequence* technique (Figure 2.10). In it the data sequence is multiplied directly − before modulation − with a spreading sequence to generate the band-spread signal. The bit rate of the spreading signal, the so-called *chip rate,* is obtained by multiplying the bit rate of the data signal by the spreading factor, which generates the desired broadening of the signal spectrum. Ideally, the spreading sequences are completely orthogonal bit sequences ("codes") with disappearing cross-correlation functions. Since such completely orthogonal sequences cannot be realized, practical systems use bit sequences from *pseudonoise* (PN) generators to spread the band [29][32]. For despreading, the signal is again multiplied with the spreading sequence at the receiver, which ideally recovers the data sequence in its original form.

Thus one can realize a code-based multiple access system. If an orthogonal family of spreading sequences is available, each subscriber can be assigned his or her own unique spreading sequence. Because of the disappearing cross-correlation of the spreading sequences, the signals of the individual subscribers can be separated in spite of being transmitted in the same frequency band at the same time.

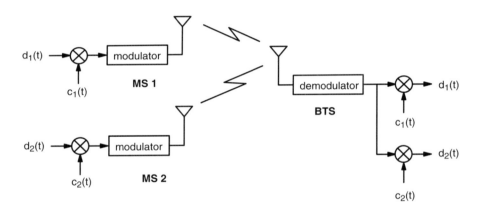

Figure 2.11: Simplified scheme of code division multiple access (uplink)

In a simplified way, this is done by multiplying the received summation signal with the respective code sequence (Figure 2.11):

$$s(t)\, c_j(t) \;=\; c_j(t) \sum_{i=1}^{n} d_i(t) c_i(t) \;=\; d_j(t)$$

$$\text{with } \; c_j(t) c_i(t) \;=\; \begin{cases} 0, i \neq j \\ 1, i = j \end{cases}$$

Thus, if direct sequence spreading is used, the procedure is called *Direct Sequence Code Division Multiple Access* (DS-CDMA).

Another possibility for spreading the band is the use of a fast frequency hopping technique. If one changes the frequency several times during one transmitted data symbol, a similar spreading effect occurs as in case of the direct sequence procedure. If the frequency hopping sequence is again controlled by orthogonal code sequences, another multiple access system can be realized, the *Frequency Hopping CDMA* (FH-CDMA).

2.3.4 Space Division Multiple Access (SDMA)

An essential property of the mobile radio channel is multipath propagation, which leads to frequency-selective fading phenomena. Furthermore, multipath propagation is the cause of another significant property of the mobile radio channel, the spatial fanning out of signals. This causes the received signal to be a summation signal, which is not only determined by the *Line of Sight* (LOS) connection but also by an undetermined number of individual paths caused by refractions, infractions, and reflections. In principle, the directions of incidence of these multipath components could therefore be distributed arbitrarily at the receiver.

Especially on the uplink from the mobile station to the base station, there is, however, in most cases a main direction of incidence (usually LOS), about which the angles of incidence of the individual signal components are scattered in a relatively narrow range. Frequently, the essential signal portion at the receiver is distributed only over an angle of a few tens of degrees. This is because base stations are installed wherever possible as free-standing units, and there are no interference centers in the immediate neighborhood.

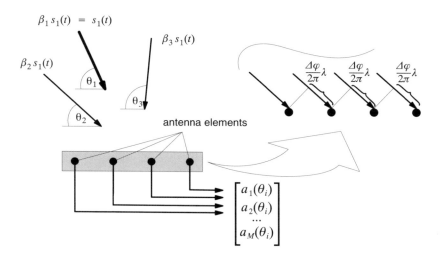

Figure 2.12: Multipath signal at an antenna array

This directional selectivity of the mobile radio channel, which exists in spite of multipath propagation, can be exploited by using array antennas. Antenna arrays generate a directional characteristic by controlling the phases of the signals from the individual antenna elements. This allows the receiver to adjust the antenna selectively to the main direction of incidence of the received signal, and conversely to transmit selectively in one direction. This principle can be illustrated easily with a simple model (Figure 2.12).

The individual multipath components $\beta_i s_1(t)$ of a transmitted signal $s_1(t)$ propagate on different paths such that the multipath components incident at an antenna under the angle θ_i differ in amplitude and phase. If one considers an array antenna with M elements ($M=4$

in Figure 2.12) and a wave front of a multipath component incident at angle θ_i on this array antenna, then the received signals at the antenna elements differ mainly in their phase — each shifted by $\Delta\varphi$ (Figure 2.12) — and amplitude.

In this way, the response of the antenna to a signal incident at angle θ_i can be characterized by the complex response vector $\vec{a}(\theta_i)$ which defines amplitude gain and phase of each antenna element relative to the first antenna element ($a_1=1$):

$$\vec{a}(\theta_i) = \begin{bmatrix} a_1(\theta_i) \\ a_2(\theta_i) \\ \cdots \\ a_M(\theta_i) \end{bmatrix} = \begin{bmatrix} 1 \\ a_2(\theta_i) \\ \cdots \\ a_M(\theta_i) \end{bmatrix}$$

The N_m multipath components ($N_m=3$ in Figure 2.12) of a signal $s_1(t)$ generate, depending on the incidence vector θ_i, a received signal vector $\vec{x_1}(t)$ which can be written with the respective antenna response vector $\vec{a}(\theta_i)$ and the signal of the ith multipath $\beta_i s_1(t)$ shifted in amplitude and phase against the direct path $s_1(t)$ as

$$\vec{x_1}(t) = \vec{a}(\theta_1) s_1(t) + \sum_{i=2}^{N_m} \vec{a}(\theta_i) \beta_i s_1(t) = \vec{a_1} s_1(t)$$

In this case, the vector $\vec{a_1}$ is also designated the spatial signature of the signal $s_1(t)$, which remains constant as long as the source of the signal does not move and the propagation conditions do not change [8]. In a multi-access situation, there are typically several sources (N_q); this yields the following result for the total signal at the array antenna: neglecting noise and interferences,

$$\vec{x}(t) = \sum_{j=1}^{N_q} \vec{a_j} s_j(t)$$

From this summation signal, the signals of the individual sources are separated by weighting the received signals of the individual antenna elements with a complex factor (weight vector $\vec{w_i}$), which yields

$$\vec{w_i}^H \vec{a_j} = \begin{cases} 0, & i \neq j \\ 1, & i = j \end{cases}$$

For the weighted summation signal [8] one gets

$$\vec{w_i}^H \vec{x}(t) = \sum_{j=1}^{N_q} \vec{w_i}^H \vec{a_j} s_j(t) = s_i(t)$$

Under ideal conditions, i.e. neglecting noise and interference, the signal $s_i(t)$ of a single source i can be separated from the summation signal of the array antenna by using an appropriate weight vector during signal processing. The determination of the respectively optimal weight vector, however, is a nontrivial and computation-intensive task. Because of the considerable processing effort and also because of the mechanical dimensions of the antenna field, array antennas are predominantly used in base stations.

So far only the receiving direction has been considered. The corresponding principles, however, can also be used for constructing the directional characteristics of the transmitter. Assume symmetric propagation conditions in the sending and receiving directions, and assume the transmitted signals $s_i(t)$ are weighted with the same weight vector $\vec{w}_i$ as the received signal, before they are transmitted through the array antenna; then one obtains the following summation signal radiated by the array antenna:

$$\vec{y}(t) = \sum_{j=1}^{N_q} \vec{w}_j s_j(t)$$

and for the signal received on the ith opposite side respectively:

$$\hat{s}_i(t) = \vec{a}_i^H \vec{y}(t) = \sum_{j=1}^{N_q} \vec{a}_i^H \vec{w}_j s_j(t) = s_i(t)$$

Thus, by using array antennas, one can separate the simultaneously received signals of spatially separated subscribers by exploiting the directional selectivity of the mobile radio channel. Because of the use of intelligent signal processing and corresponding control algorithms, such systems are also known as systems with intelligent antennas.

The directional characteristics of the array antenna can be controlled adaptively such that a signal is only received or transmitted in exactly the spatial segment where a certain mobile station is currently staying. On one hand, one can thus reduce cochannel interference in other cells, and on the other hand, the sensitivity against interference can be reduced in the current cell. Furthermore, because of the spatial separation, physical channels in a cell can be reused , and the lobes of the antenna diagram can adaptively follow the movement of mobile stations. In this case, yet another multiple access technique (Figure 2.13) is defined and known as *Space Division Multiple Access* (SDMA).

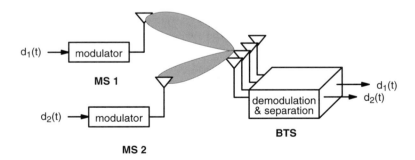

Figure 2.13: Schematic representation of spatial multiple access (uplink)

SDMA systems are currently the subject of intensive research. The SDMA technique can be combined with each of the other multiple access techniques (FDMA, TDMA, CDMA). This enables intracellular spatial channel reuse, which again increases the network capacity [37]. This is especially attractive for existing networks which can use an intelligent implementation of SDMA by selectively upgrading base stations with array antennas, appropriate signal processing, and respective control protocols.

2.4   Cellular Technology

Because of the very limited frequency bands, a mobile radio network has only a relatively small number of speech channels available. For example, the GSM system has an allocation of 25 MHz bandwidth in the 900 MHz frequency range, which amounts to a maximum of 125 frequency channels each with a carrier bandwidth of 200 kHz. Within an eightfold timemultiplex for each carrier, a maximum of 1000 channels can be realized. This number is further reduced by guardbands in the frequency spectrum and the overhead required for signaling (Chapter 5). In order to be able to serve several 100 000 or millions of subscribers in spite of this limitation, frequencies must be spatially reused, i.e. deployed repeatedly in a geographic area. In this way, services can be offered with a cost-effective subscriber density and acceptable blocking probability.

2.4.1   Fundamental Definitions

This spatial frequency reuse concept led to the development of cellular technology, which allowed a significant improvement in the economic use of frequencies. The essential characteristics of this procedure are as follows:

- The area to be covered is subdivided into cells (radio zones). For easier manipulation, these cells are modeled in a simplified way as hexagons (Figure 2.14). Most models show the base station in the middle of the cell.

- Each cell i receives a subset of the frequencies fb_i from the total set (bundle) assigned to the respective mobile radio network. Two neighboring cells must never use the same frequencies, since this would lead to severe cochannel interference from the immediately adjacent cells.

- Only at distance D (the frequency reuse distance) can a frequency from the set fb_i be reused (Figure 2.4), i.e. cells with distance D to cell i are assigned one or all of the frequencies from the set fb_i belonging to cell i. If D is chosen sufficiently large, the cochannel interference remains small enough not to affect speech quality.

- When moving from one cell to another during an ongoing conversation, an automatic channel / frequency change occurs (handover), which maintains an active speech connection over cell boundaries.

The spatial repetition of frequencies is done in a regular systematic way, i.e. each cell with the frequency allocation fb_i (or one of its frequencies) sees its neighbors with the same frequencies again at a distance D (Figure 2.14). Therefore there exist exactly six such next neighbor cells. Independent of form and size of the cells − not only in the hexagon model − the first ring in the frequency set contains six cochannel cells (see also Figure 2.15).

2.4.2 Signal-to-Noise Ratio

The interference caused by neighboring cells is measured as the signal-to-noise-ratio:

$$W = \frac{\text{useful signal}}{\text{disturbing signal}} = \frac{\text{useful signal}}{\text{neighbor cell interference} + \text{noise}}$$

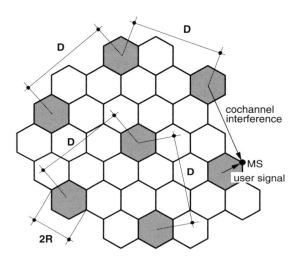

Figure 2.14: Model of a cellular network with frequency reuse

This ratio of the useful signal to the interfering signal is usually measured in decibels (dB) and called the *Signal-to-Noise Ratio* (SNR). The intensity of the interference is essentially a function of cochannel interference depending on the frequency reuse distance D. From the viewpoint of a mobile station, the cochannel interference is caused by base stations at distance D from the current base station. A worst-case estimate for the signal-to-noise ratio W of a mobile station at the border of the covered area at distance R from the base station can be obtained, subject to propagation losses, by assuming that all six neighboring interfering transmitters operate at the same power and are approximately equally far apart (distance D large against cell radius R) [25]:

$$W = \frac{P_0 R^{-\gamma}}{\sum_{i=1}^{6} P_i + N} \approx \frac{P_0 R^{-\gamma}}{\sum_{i=1}^{6} P_0 D^{-\gamma} + N} = \frac{P_0 R^{-\gamma}}{6 P_0 D^{-\gamma} + N}$$

By neglecting the noise N we obtain the following approximation for the carrier-to-interference ratio C/I:

$$W \approx \frac{C}{I} = \frac{R^{-\gamma}}{6 D^{-\gamma}} = \frac{1}{6}\left(\frac{R}{D}\right)^{-\gamma}$$

Therefore the signal-to-noise ratio depends essentially on the ratio of the cell radius R to the frequency reuse distance D. From these considerations it follows that for a desired or needed signal-to-noise ratio W at a given cell radius, one must choose a minimum distance for the frequency reuse, above which the cochannel interference fall below the required threshold.

2.4.3 Formation of Clusters

The regular repetition of frequencies results in a clustering of cells. The clusters generated in this way can comprise the whole frequency band. In this case all of the frequencies in the available spectrum are used within a cluster. The size of a cluster is characterized by the number of cells per cluster k, which determines the frequency reuse distance D. Figure 2.15 shows some examples of clusters. The numbers designate the respective frequency sets fb_i used within the single cells.

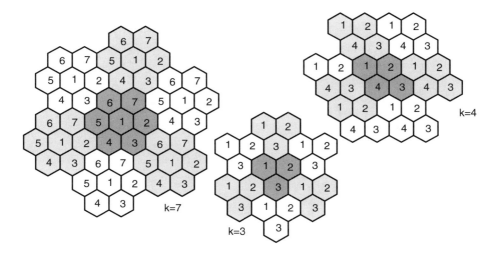

Figure 2.15: Frequency reuse and cluster formation

For each cluster the following holds:

- A cluster can contain all the frequencies of a mobile radio system.
- Within a cluster, no frequency can be reused. The frequencies of a set fb_i may be reused at the earliest in the neighboring cluster.
- The larger a cluster, the larger the frequency reuse distance and the larger the signal-to-noise ratio. However, the larger the values of k, the smaller the number of channels and the number of active subscribers per cell.

The frequency reuse distance D can be derived geometrically from the hexagon model depending on k and the cell radius R:

$$D = L\sqrt{3k}$$

The signal-to-noise ratio W [25] is then

$$W = \frac{R^{-\gamma}}{6\,D^{-\gamma}} = \frac{R^{-\gamma}}{6\left(R\sqrt{3k}\right)^{-\gamma}} = \frac{1}{6}(3k)^{\gamma/2}$$

According to measurements one can assume that, for good speech understandability, a *Carrier-to-Interference Ratio* (CIR) of about 18 dB is sufficient. Assuming an approximate propagation coefficient of $\gamma=4$, this yields the minimum cluster size

$$10 \, \log W \geq 18 \, \text{dB}; \; W \geq 63.1 \Rightarrow D \approx 4,.4\,R$$

$$\frac{1}{6}(3k)^{\gamma/2} = W \geq 63.1 \Rightarrow k \geq 6.5 \Rightarrow k = 7$$

These values are also confirmed by computer simulations, which have shown that for W = 18 dB a reuse distance $D = 4.6R$ becomes necessary [25]. In practically implemented networks, one can find other cluster sizes, e.g. $k=3$ and $k=12$. A carrier-to-interference ratio of 15 dB is considered a conservative value for network engineering.

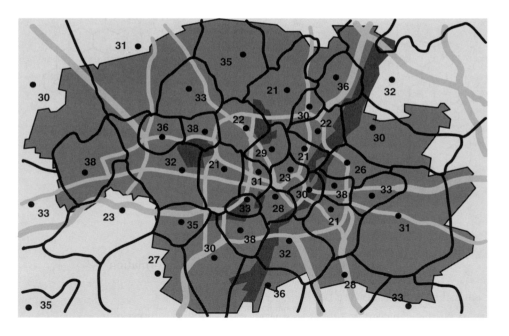

Figure 2.16: Cell structure of a real network

The cellular models mentioned so far are very idealized for illustration and analysis. In reality, cells are neither circular nor hexagonal; rather they possess very irregular forms and sizes because of variable propagation conditions. An example of a possible cellular plan for a real network is shown in Figure 2.16, where one can easily recognize the individual cells with the assigned channels and the frequency reuse. Especially obvious are the different cell sizes, which depend on whether it is an urban, suburban, or rural area. Figure 2.16 gives an impression of the approximate contours of equal signal power around the individual base stations. In spite of this representation, the precise fitting of signal power contours remains an idealization. The cell boundaries are after all blurred and defined by local thresholds, beyond which the neighboring base station's signal is received stronger than the current one.

2.4.4   Traffic Capacity and Traffic Engineering

As already mentioned, the number of channels and thus the maximal traffic capacity per cell depends on the cluster size k. The following relation holds:

$$n_F = \frac{B_t}{B_c k} ,$$

where

n_F = number of frequencies per cell
B_t = total bandwidth of the system
B_c = bandwidth of one channel

The number of channels per cell in FDMA systems equals the number of frequency channels resulting from the channel and system bandwidth:

$$n = n_F$$

The number of channels per cell in a TDMA system is the number of frequency channels multiplied by the number of time slots per channel (frame size):

$$n = m n_F$$

where

m = number of time slots per frame

A cell can be modeled as a traffic-theoretical loss system with n servers (channels), assuming a call arrival process with exponentially distributed interarrival times (Poisson process), and another Poisson process as a server process. Arrival and server processes are also called Markov processes, hence such a system is known as an M/M/n loss system [1]. For a given blocking probability B, a cell serves a maximum offered load A_{max} during the busy hour:

$$A_{max} = f(B, n) = \lambda_{max} T_m ,$$

where

λ_{max} = busy hour call attempts (BHCA)
T_m = mean call holding time

The relation between offered load A and blocking probability B with the total number of channels n is given by the Erlang blocking formula (see [1] [2] for more details and traffic tables):

$$B = \frac{A^n/n!}{\sum_{i=0}^{n} A^i/i!}$$

However, these approximations are valid only for macrocellular environments, in which the number of users per cell is sufficiently large with regard to the number of available channels, such that the call arrival rate may be considered as approximately constant. For micro- and picocellular systems these assumptions don't usually hold any more. The traffic-theoretical dimensioning must here be done with Engset models, since the number of participants does not differ very much from the number of available channels. This results in a call arrival rate that is no longer constant. The probability that all channels are busy results from the number of users M per cell and the offer a of a free source at:

$$P_n = \frac{\binom{M}{n} a^n}{\sum_{i=0}^{n} \binom{M}{i} a^i}$$

In this case, the probability that a call arrives when no free channels are available (blocking probability) is

$$P_B = \frac{\binom{M-1}{n} a^n}{\sum_{i=0}^{n} \binom{M-1}{i} a^i}$$

For $M \rightarrow \infty$, the Engset blocking formula becomes the Erlang blocking formula.

3 System Architecture and Addressing

3.1 General Description

GSM networks are structured hierarchically (Figure 3.1). Essentially they consist of at least one administrative region, which is assigned to a mobile switching center. Each administrative region is made up of at least one *Location Area* (LA). Frequently, the LA is also called the visited area. An LA consists of several cell groups. Each cell group is assigned to a base station controller (BSC). Therefore for each LA there exists at least one BSC, but cells of one BSC may belong to different LAs.

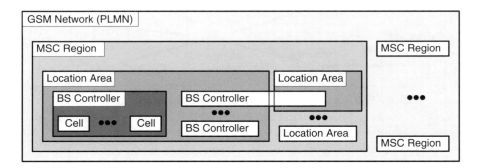

Figure 3.1: GSM system hierarchy

The effort required to cover all of Germany is illustrated in Table 3.1 for the time shortly after the start of the two German GSM900 networks (the networks D1 and D2) and for the time of the planned completion of the networks. This effort is similar for both network operators. The exact partitioning of the service area into cells and their organization or administration with regard to LAs, BSSs, and MSCs is, however, not uniquely determined and is left to the respective network operator who thus has many possibilities for optimization.

Table 3.1: Evolution of a GSM network in Germany

	Cells	BSC	MSC
End of 1992	approx. 500	7	7
Total completion	approx. 7000	20 – 50	20 – 50

Figure 3.2 shows the system architecture of a GSM *Public Land Mobile Network* (PLMN) with essential components. The hierarchical construction of the GSM infrastructure becomes evident again. The cell is formed by the radio area coverage of a *Base Transceiving Station* (BTS). Several base stations together are controlled by one *Base Station Controller* (BSC). The combined traffic of the mobile stations in their respective cells is routed through a switch, the *Mobile Switching Center* (MSC). Conversations originating from or terminating in the fixed network are handled by a dedicated *Gateway Mobile Switching Center* (GMSC). Operation and maintenance are organized from a central place, the *Operation and Maintenance Center* (OMC). Several databases are available for call control and network management: the *Home Location Register* (HLR), the *Visitor Location Register* (VLR), the *Authentication Center* (AUC), and the *Equipment Identity Register* (EIR).

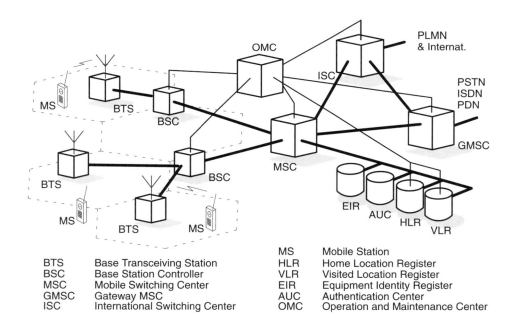

BTS	Base Transceiving Station	MS	Mobile Station
BSC	Base Station Controller	HLR	Home Location Register
MSC	Mobile Switching Center	VLR	Visited Location Register
GMSC	Gateway MSC	EIR	Equipment Identity Register
ISC	International Switching Center	AUC	Authentication Center
		OMC	Operation and Maintenance Center

Figure 3.2: GSM system architecture with essential components

For all subscribers registered with a network operator, permanent data such as the service profile as well as temporary data such as the current location are stored in the HLR. In case of a call to a mobile station, the HLR is always first interrogated, to determine the mobile station's current location. A VLR is responsible for a group of LAs and stores the data of those subscribers which are currently in its area of responsibility. This includes parts of the permanent subscriber data which have been transmitted from the HLR to the VLR for faster access. But the VLR may also assign and store local data such as a temporary identification. The AUC generates and stores security-related data such as keys used for authentication and encryption, whereas the EIR registers equipment data rather than subscriber data.

3.2   Addresses and Identifiers

GSM distinguishes explicitly between user and equipment and deals with them separately. According to this concept, which was introduced with digital mobile networks, mobile equipment and users each receive their own internationally unique identifiers. The user identity is associated with a mobile station by means of a personal chip card, the *Subscriber Identity Module* (SIM). This SIM is in form of a credit card which is portable and therefore transferable between mobile stations. It enables distinction between equipment mobility and subscriber mobility. The subscriber can register into the locally available network with his or her SIM chipcard on different pieces of mobile station equipment, or the SIM card could be used as a normal telephone card in the fixed telephone network. However, he or she cannot receive calls on fixed network ports, but further development of the fixed networks as well as convergence of fixed and mobile networks could make this possible too. In that case, a mobile subscriber could register at an arbitrary ISDN telephone and be able to accept calls.

In addition, GSM distinguishes between subscriber identity and telephone number. This leaves some leeway for development of future services when each subscriber may be called personally, independent of reachability or type of connection (mobile or fixed). Besides the personal identifier, each GSM subscriber is assigned one or several ISDN numbers.

Besides telephone numbers, subscriber and equipment identifiers several other identifiers have been defined; they are needed for the management of subscriber mobility and for addressing all the remaining network elements.

The most important addresses and identifiers will be presented in the following.

3.2.1   International Mobile Station Equipment Identity

The *International Mobile Station Equipment Identity* (IMEI) uniquely identifies mobile equipment internationally. It is a kind of serial number. The IMEI is allocated by the equipment manufacturer and registered by the network operator who stores it in the *Equipment Identity Register* (EIR).

By means of the IMEI one recognizes obsolete, stolen, or nonfunctional equipment and, for example, can deny service. For this purpose, the IMEI is assigned to one or more of three categories within the EIR :

- The *White List* is a register of all equipment.

- The *Black List* contains all suspended equipment. This list is periodically exchanged among the network operators.
- Optionally, an operator may carry a *Gray List*, in which malfunctioning equipment or equipment with obsolete software versions is registered. Service to such equipment is not suspended, and it has network access, but it is reported to the operating personnel, to take appropriate measures.

The IMEI is usually requested from the network at registration, but it can be requested repeatedly.

The IMEI is a hierarchical address. It consists of the following parts:

- *Type Approval Code* (TAC): 6 decimal places centrally assigned
- *Final Assembly Code* (FAC): 6 decimal places assigned by the manufacturer
- *Serial Number* (SNR): 6 decimal places assigned by the manufacturer
- *Spare* (SP): 1 decimal place

Thus IMEI = TAC + FAC + SNR + SP. It uniquely characterizes a mobile station and gives clues about the manufacturer and the date of manufacturing.

3.2.2   International Mobile Subscriber Identity

When registering for service with a mobile network operator, each subscriber receives a unique identifier, the *International Mobile Subscriber Identity* (IMSI). This IMSI is stored in the subscriber identity module (SIM); see Section 3.3.1. A mobile station can only be operated, if a SIM with a valid IMSI is inserted into equipment with a valid IMEI, since this is the only way to correctly bill the appropriate subscriber.

The IMSI also consists of several parts:

- *Mobile Country Code* (MCC): 3 decimal digits, internationally standardized
- *Mobile Network Code* (MNC): 2 decimal digits, for unique identification of mobile networks within a country
- *Mobile Subscriber Identification Number* (MSIN): maximum 10 decimal digits, identification number of the subscriber in his mobile home network

The IMSI is a GSM-specific addressing concept in contrast to the ISDN numbering plan. A 3-digit MCC was assigned to each of the GSM countries, and 2-digit MNCs were assigned within countries (e.g. 262 as MCC for Germany and MNC 01, 02, and 03 for the networks known as D1-Telekom, D2-Privat, and E-Plus, respectively). Subscriber identification therefore uses a maximum of 15 decimal digits, and IMSI=MCC+MNC+MSIN. Whereas the MCC is defined internationally, the *National Mobile Subscriber Identity* (NMSI, NMSI = MNC + MSIN) is assigned by the operator of the home PLMN.

3.2.3   Mobile Subscriber ISDN Number

The "real telephone number" of a mobile station is the *Mobile Subscriber ISDN Number* (MSISDN). It is assigned to the subscriber (his or her SIM, respectively), such that a mobile station set can have several MSISDNs depending on the SIM. With this concept, GSM is the first mobile system to distinguish between subscriber identity and number to call. The separation of call number MSISDN and subscriber identity IMSI primarily serves to protect the confidentiality of the IMSI. In contrast to the MSISDN, the IMSI need not be made public. With this separation, one cannot derive the subscriber identity from the MSISDN, unless the association of IMSI and MSISDN as stored in the HLR has been made public. It is the rule that the IMSI used for subscriber identification is not known, and thus the faking of a false identity is significantly more difficult.

Beyond that, a subscriber can hold several MSISDNs for selection of different services. Each MSISDN of a subscriber is reserved for specific service (voice, data, fax, etc.). In order to realize this service, service-specific resources have to be activated in the mobile station as well as in the network. The service desired and the resources needed for the specific call can be recognized from the MSISDN. Thus an automatic activation of service-specific resources is already possible during the setup of a connection. The MSISDN categories follow the international ISDN numbering plan and therefore have the following structure:

- *Country Code* (CC): up to 3 decimal places
- *National Destination Code* (NDC): typically 2 to 3 decimal places
- *Subscriber Number* (SN): maximal 10 decimal places

The country identifiers CC are internationally standardized, complying to the ITU-T E.164 series [16]. There are country codes with one, two, or three digits; e.g. the country code for the USA is 1, for the UK it is 44, and for Finland it is 358. The national operator or regulatory administration assigns the NDC as well as the subscriber number SN, which may have variable length. The NDC of the mobile networks in Germany have three digits (171, 172, 173). The subscriber number is the concatenation MSISDN=CC+NDC+SN and thus has a maximum of 15 decimal digits. It is stored centrally in the HLR.

3.2.4   Mobile Station Roaming Number

The *Mobile Station Roaming Number* (MSRN) is a temporary location-dependent ISDN number. It is assigned by the locally responsible VLR to each mobile station in its area. Calls are routed to the MS by using the MSRN. On request the MSRN is passed from the HLR to the GMSC. The MSRN has the same structure as the MSISDN:

- *Country Code* (CC): of the currently visited network
- *National Destination Code* (NDC): of the visited network
- *Subscriber Number* (SN): in the current mobile network

The components CC and NDC are determined by the visited network and depend on the current location. The SN is assigned by the current VLR and is unique within the mobile network. The assignment of an MSRN is done in such a way that the currently responsible switching node MSC in the visited network (CC+NDC) can be determined from the subscriber number, which allows routing decisions to be made.

The MSRN can be assigned in two ways by the VLR: either at each registration when the MS enters a new location area LA or each time when the HLR requests it for setting up a connection for incoming calls to the mobile station.

In the first case, the MSRN is also passed on from the VLR to the HLR where it is stored for routing. In the case of an incoming call, the MSRN is first requested from the HLR of this mobile station. This way the currently responsible MSC can be determined, and the call can be routed to this switching node. Additional localization information can be obtained there from the responsible VLR.

In the second case, the MSRN cannot be stored in the HLR, since it is only assigned at the time of call setup. Therefore the address of the current VLR must be stored in the tables of the HLR. Once routing information is requested from the HLR, the HLR itself goes to the current VLR and uses a unique subscriber identification (IMSI and MSISDN) to request a valid roaming number MSRN. This allows further routing of the call.

3.2.5 Location Area Identity

Each *Location Area* (LA) of a PLMN has its own identifier. The *Location Area ID* (LAI) is also structured hierarchically and internationally unique (Section 3.2.2), with LAI again consisting of an internationally standardized part and an operator-dependent part:

- *Country Code* (CC): 3 decimal digits
- *Mobile Network Code* (MNC): 2 decimal digits
- *Location Area Code* (LAC): maximum 5 decimal digits, or maximum twice 8 bits, coded in hexadecimal (LAC $<$ FFFF$_{hex}$)

This LAI is broadcast regularly by the base station on the *Broadcast Control Channel* (BCCH). Thus each cell is identified uniquely on the radio channel as belonging to an LA, and each MS can determine its current location through the LAI. If the LAI "heard" by the MS changes, the MS notices this LA change and requests the updating of its location information in the VLR and HLR (location update). The significance for GSM networks is that the mobile station itself rather than the network is responsible for monitoring the local conditions of signal reception, to select the base station that can be received best, and to register with the VLR of that LA which the current base station belongs to.

The LAI is requested from the VLR if the connection for an incoming call has been routed to the current MSC using the MSRN. This determines the precise location of the mobile station where the mobile can be subsequently paged. When the mobile station answers, the exact cell and therefore also the base station become known; this information can then be used to switch the call through.

3.2.6 Temporary Mobile Subscriber Identity

The VLR responsible for the current location of a subscriber can assign a *Temporary Mobile Subscriber Identity* (TMSI) which has only local significance in the area handled by the VLR. It is used in place of the IMSI for the definite identification and addressing of the mobile station. This way nobody can determine the identity of the subscriber by listening to the radio channel, since this TMSI is only assigned during the mobile station's presence in the area of one VLR, and can even be changed during this period (ID hopping). The mobile station stores the TMSI on the SIM card. The TMSI is stored on the network side only in the VLR and is not passed to the HLR.

A TMSI may therefore be assigned in an operator-specific way; it can consist of up to 4×8 bits, but the value FFFF FFFF$_{hex}$ is excluded, because the SIM marks empty fields internally with logical 1.

Together with the current location area, a TMSI allows a subscriber to be identified uniquely, i.e. for the ongoing communication the IMSI is replaced by the 2-tuple (TMSI, LAI).

3.2.7   Local Mobile Subscriber Identity

The VLR can assign to a mobile station in its area an additional searching key to accelerate database access; this is the *Local Mobile Station Identity* (LMSI). The LMSI is assigned when the mobile station registers with the VLR and is also sent to the HLR. The LMSI is not used any further by the HLR, but each time messages are sent to the VLR concerning a mobile station, the LMSI is added, so the VLR can use the short searching key for transactions concerning this MS. This kind of additional identification is only used when the MSRN is newly assigned with each call. In this case, fast processing is very important to achieve short times for call setup.

Like the TMSI, an LMSI is also assigned in an operator-specific way, and it is only unique within the administrative area of a VLR. An LMSI consists of four octets (4×8 bits).

3.2.8   Cell Identifier

Within an LA, the individual cells are uniquely identified with a *Cell Identifier* (CI), maximum 2×8 bits. Together with the *Global Cell Identity* (LAI+CI), cells are thus also internationally defined in a unique way.

3.2.9   Base Transceiver Station Identity Code

In order to distinguish neighboring base stations, these receive a unique *Base Transceiver Station Identity Code* (BSIC) which consists of two components:
- *Network Color Code* (NCC): color code within a PLMN (3 bits)
- *Base Station Color Code* (BCC): BS color code (3 bits)

The BSIC is broadcast periodically by the base station on a *Broadcast Channel*, the *Synchronization Channel*. Directly adjacent PLMN (and BS) must have different color codes.

3.2.10 Identification of MSCs and Location Registers

MSCs and location registers (HLR, VLR) are addressed with ISDN numbers. In addition, they may have a *Signaling Point Code* (SPC) within a PLMN, which can be used to address them uniquely within the Signaling System Number 7 network (SS#7).

The number of the VLR in whose area a mobile station is currently roaming must be stored in the HLR data for this MS, if the MSRN distribution is on a call-by-call basis (Section 3.2.4); thus the MSRN can be requested for incoming calls and the call can be switched through to the MS.

3.3   System Architecture

A GSM system has two major components: the fixed installed infrastructure (the network in the proper sense) and the mobile subscribers, which use the services of the network and communicate over the radio interface (air interface).

The fixed installed GSM network can again be subdivided into three subnetworks: the radio network, the mobile switching network, and the management network [7]. These subnetworks are called *subsystems* in the GSM standard. The respective three subsystems are the *Base Station Subsystem* (BSS), the *Switching and Management Subsystem* (SMSS), and the *Operation and Maintenance Subsystem* (OMSS).

3.3.1   Mobile Station

Mobile stations (MS) are pieces of equipment which are used by mobile service subscribers for access to services. They consist of two major components: the *Mobile Equipment* and the *Subscriber Identity Module* (SIM). Only the SIM of a subscriber turns a piece of mobile equipment into a complete mobile station with network usage privileges, which can be used to make calls or receive calls. The SIM can be a fixed installed chip (*plug-in SIM*) or an exchangeable SIM card. In addition to the equipment identifier IMEI, the mobile station has subscriber identification and call number (IMSI and MSISDN) as subscriber-dependent data. Thus GSM mobile stations are personalized with the SIM card (Figure 3.3).

Figure 3.3: Mobile equipment personalization with the SIM

This modern concept of the SIM used consistently for the first time in GSM achieved on one hand the separation of user mobility from equipment mobility. This enables international roaming independent of mobile equipment and network technology, provided the interface between SIM and end terminal is standardized. On the other hand, the SIM can assume substantially more tasks than the personalization of mobile stations with IMSI and MSISDN. All the cryptographic algorithms to be kept confidential are realized on the SIM, which implements important functions for the authentication and user data encryption based on the subscriber identity IMSI and secret keys. Beyond that, the SIM can store short messages and charging information, and it has a telephone book function and short list of call numbers storing names and telephone numbers for efficient and fast number selection. These functions in particular contribute to a genuine personalization of a mobile terminal, since the subscriber can use his or her normal "environment" plus telephone list and short message archive with any piece of mobile equipment. Besides subscriber-specific data, the SIM can also store network-specific data, e.g. lists of BCCH carrier frequencies used by the network to broadcast system information periodically, or also the current LAI. Use of the SIM and thus of the whole MS can be protected with a PIN against unauthorized access.

3.3.2   Radio Network (BSS)

Figure 3.4 shows the components of the GSM radio network (BSS). A GSM cell is expanded around the radio area of a *Base Transceiver Station* (BTS); *transmitter+receiver=transceiver*. The BTS provides the radio channels for signaling and user data traffic in this cell. A BTS is therefore on the network side of the air interface of a GSM PLMN. Besides the high-frequency part (transmitter and receiver equipment) it contains only a few components for signal and protocol processing. For example, error protection coding and coding for the radio channel are performed in the BTS, and the link level protocol LAPDm for signaling on the radio path is terminated here. In order to keep the base stations small, the essential control and protocol intelligence entities reside in the *Base Station Controller* (BSC). For example, the handover protocol is executed in the BSC. BTS and BSC together form the *Base Station Subsystem* (BSS). Several BTSs can be controlled together by one BSC (Figure 3.1). Each BTS is allocated a set of frequency channels, the *Cell Allocation* (CA).

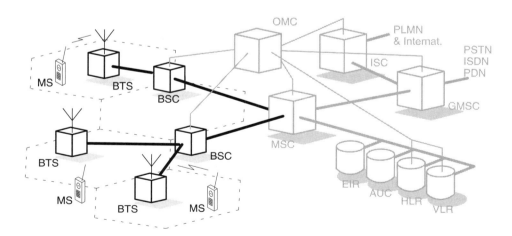

Figure 3.4: Components of the GSM radio network

Two kinds of channels are provided at the radio interface: traffic channels and signaling channels. Traffic channels are further subdivided into full-rate channels and half-rate channels. For the traffic channels, the BSS comprises essentially all the functions of OSI Layer 1.

3.3.3   Mobile Switching Network (MSS)

The *Mobile Switching and Management Subsystem* (SMSS) consists of the mobile switching centers and the databases which store the data required for routing and service provision (Figure 3.5). These components and their functions will be presented briefly in the following. Later sections will discuss more details.

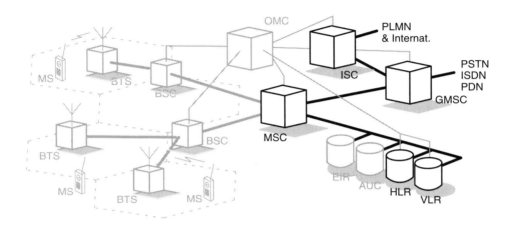

Figure 3.5: Components of the GSM mobile switching network

3.3.3.1   Mobile Switching Center

The switching node of a GSM PLMN is the *Mobile Switching Center* (MSC). The MSC performs all the switching functions of a fixed-network switching node, e.g. routing path search, signal routing, and service feature processing. The main difference between an ISDN switch and an MSC is that the MSC also has to consider the allocation and administration of radio resources and the mobility of the subscribers. The MSC therefore has to provide additional functions for location registration of subscribers and for the handover of a connection in case of changing from cell to cell. A PLMN can have several MSCs with each being responsible for a part of the *Service Area*. The BSCs of a BSS are subordinated to a single MSC.

Dedicated *Gateway MSCs* (GMSCs) are available to pass voice traffic between fixed networks and mobile networks. If due to the inability to interrogate the HLR, the fixed network is unable to connect an incoming call to the local MSC, it routes the connection to the next GMSC. This GMSC requests the routing information from the HLR and routes the connection to the local MSC in whose area the mobile station is currently staying. Connections to other mobile or international networks are mostly routed over the *International Switching Center* (ISC) of the respective country.

Associated with an MSC is a functional unit enabling the interworking of a PLMN and the fixed networks (PSTN, ISDN, PDN). This *Interworking Function* (IWF) performs a variety of functions depending on the service and the respective fixed network. It is needed to map the protocols of the PLMN onto those of the respective fixed network. In cases of compatible service implementation in both networks, the IWF has no functions to perform.

3.3.3.2   Home and Visitor Registers

A GSM PLMN has several databases. Two functional units are defined for the registration of subscribers and their current location: the *Home Location Register* (HLR) and the *Visitor Location Register* (VLR). In general, there is one central HLR per PLMN and one VLR for each MSC. This organization depends on the number of subscribers, the processing and storage capacity of the switches and the structure of the network.

The HLR is the home register which has entries for every subscriber and every mobile ISDN number which has its "home" in the respective network. It stores all permanent subscriber data and the relevant temporary data of all subscribers permanently registered in the HLR. Besides the fixed entries like service subscriptions and permissions, the stored data also contains a hint to the current location of the mobile station (Table 3.2). The HLR is needed as the central register for the routing to the subscribers for which it has administrative responsibility. The HLR has no direct control over an MSC. All administrative activities concerning a subscriber are performed in the databases of the HLR.

The VLR as visitor register stores the data of all mobile stations which are currently staying in the administrative area of the associated MSC. A VLR can be responsible for the areas of one or more MSCs. Mobile stations can be roaming freely, and therefore, depending on their current location, they may be registered in one of the VLRs of their home network or in a VLR of a "foreign" network, if there is a roaming agreement between both network operators. For this purpose, a mobile station has to start a registration procedure when it enters a location area LA. The responsible MSC passes the identity of the MS and its current LAI to the VLR, which enters these values into its database and thus registers the MS. If the mobile station has not been registered with this VLR, the HLR is informed about the current location of the MS. In this process, information is passed to the HLR, which enables routing of incoming calls to this station.

3.3.4   Operation and Maintenance (OMSS)

3.3.4.1   Network Monitoring and Maintenance

The ongoing network operation is controlled and maintained by the *Operation and Maintenance Subsystem* (OMSS). Network control functions are monitored and initiated from an *Operation and Maintenance Center* (OMC). Here are some of its functions:

- Administration and commercial operation (subscribers, end terminals, charging, statistics)
- Security management
- Network configuration, operation, performance management
- Maintenance tasks

Management of the network can be centralized in one or more *Network Management Centers* (NMC). The operation and maintenance functions are based on the concept of the *Telecommunication Management Network* (TMN) which is standardized in the ITU-T series M.30. The OMSS components are shown in Figure 3.6.

3.3.4.2   User Authentication and Equipment Registration

Two more databases are defined in GSM besides the HLR and VLR. They are responsible for various aspects of system security. System security of GSM networks is based primarily on the verification of equipment and subscriber identity; therefore the databases serve for subscriber identification and authentication and for equipment registration. Confidential data and keys are stored or generated in the *Authentication Center* (AUC). The keys serve for user authentication and authorize the respective service access. The *Equipment Identity Register* (EIR) stores the serial numbers (supplied by the manufacturer) of the terminals (IMEI), which makes it possible to check for mobile stations with obsolete software or to block service access for mobile stations reported as stolen.

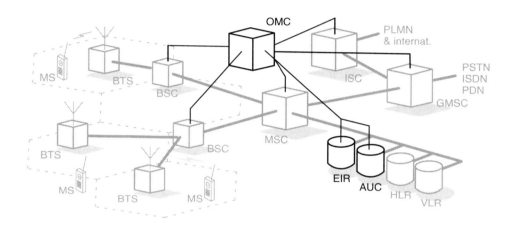

Figure 3.6: Components of the GSM OMSS

3.4   Subscriber Data in GSM

Besides data of the address type, the most important subscriber data of any communication network, a whole series of other service- and contract-specific data exists in GSM networks. Addresses serve to identify, authenticate, and localize subscribers, or switch connections to subscribers. Service-specific data is used to parametrize and personalize supplementary services. Finally, contracts with subscribers can define different service levels, e.g. booking of special supplementary services or subscriptions to data or teleservices. The contents of such contracts are of course stored in appropriate data structures in order to enable correct realization or provision of these services.

The association of the most important identifiers and their storage locations is summarized again in Figure 3.7. Subscriber-related addresses are stored on the subscriber identity module (SIM) and in the HLR and VLR as well. These data (IMSI, MSISDN, TMSI, MSRN) serve to address, identify, and localize a subscriber or a mobile station. Whereas IMSI and MSISDN are permanent data items, TMSI and MSRN are temporary values, which change according to the current location of the subscriber. Of the other data items defined for user or network equipment elements like IMEI, LAI, or SPCs only some are used (LAI, SPC) for localizing or routing. IMEI and BSIC/CI hold a special position by being used only for identification of network elements.

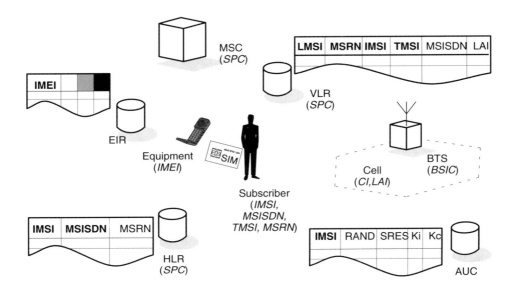

Figure 3.7: Overview of addresses and pertinent databases

Security-relevant data of a subscriber is stored in the AUC, which also calculates identifiers and keys for cryptographic processing functions. Each set of data in the AUC contains the IMSI of the subscriber as a search key. For identification and authentication of a subscriber, the AUC stores the subscriber's secret key Ki from which a pair of keys RAND/SRES are precalculated and stored. Once an authentication request occurs, this pair of keys is requested by the VLR to conduct the identification/authentication process proper. The key Kc for user data encryption on the radio channel is also calculated in advance in the AUC from the secret key Ki and is requested by the VLR at connection setup.

Table 3.2: Mobile subscriber data in the HLR

Subscriber and subscription data	Tracking and routing information
International Mobile Subscriber Identity IMSI	Mobile Station Roaming Number MSRN
International Mobile Subscriber ISDN Number MSISDN	Current VLR address (if available)
Bearer and telesevice subscriptions	Current MSC address (if available)
Sevice restrictions, e.g. roaming restrictions	Local Mobile Subscriber Identity LMSI (if available)
Parameters for additional services	
Information on the subscriber's equipment (if available)	
Authentication data (subject to implementation)	

Further data about the subscriber and his or her contractual agreement with the service provider is presented in Table 3.2 and Table 3.3. Above all, the HLR contains the permanent data about the subscriber's contractual relationship, e.g. information about subscribed to bearer and teleservices (data, fax, etc.), service restrictions, and parameters for supplementary services. Beyond that, the registers also contain information about equipment used by the subscriber (IMEI). Depending on the implementation of the authentication center AUC and the security mechanisms, data and keys used for subscriber authentication and encryption can also be stored there.

The search keys used for retrieving subscriber information such as IMSI, MSISDN, MSRN, TMSI and LMSI, from a data bank are indicated either in boldface (Figure 3.7) or in italics (Tables 3.2 and 3.3).

Table 3.3: Mobile subscriber data in the VLR

Subscriber and subscription data	Tracking and routing information
International Mobile Subscriber Identity IMSI	*Mobile Station Roaming Number MSRN*
International Mobile Subscriber ISDN Number MSISDN	*Temporary Mobile Station Identity TMSI*
Parameters for supplementary services	*Local Mobile Subscriber Identity LMSI (if available)*
Information on subscriber-used equipment (if availiable)	Location Area Identity LAI of LA, where MS was registered (used for paging and call setup)
Authentication data (subject to implementation)	

3.5   PLMN Configurations and Interfaces

The fixed connections for transport of signaling and user data in a GSM PLMN (Figure 3.8) are standard transmission lines. Within the SMSS, lines with a transmission rate of 2 Mbit/s (or 1.544 Mbit/s in North America) are typically used (fixed lines, mostly microwave links or leased lines). The BSS uses mostly 64 kbit/s lines. Signaling has two fundamentally different parts: GSM-specific signaling within the BSS, including the air interface, and signaling within the SMSS and with other PLMN in conformity with Signaling System Number 7 (SS#7). User data connections are processed with an SS#7 protocol for signaling between network nodes, the ISDN User Part (ISUP). For the mobile network specific signaling, MSC, HLR, and VLR hold extensions of SS#7, the so-called *Mobile Application Part* (MAP). Signaling between MSC and BSS uses the *Base Station System Application Part* (BSSAP). Within the BSS and at the air interface, signaling is mobile-specific, i.e. no SS#7 protocol is used here for signaling transport.

An MSC which needs to obtain data about a mobile station staying in its administrative area, requests the data from the VLR responsible for this area over the B interface. Conversely, the MSC forwards to this VLR any data generated at location updates by mobile stations. If the subscriber reconfigures special service features or activates supplementary services, the VLR is also informed first, which then updates the HLR.

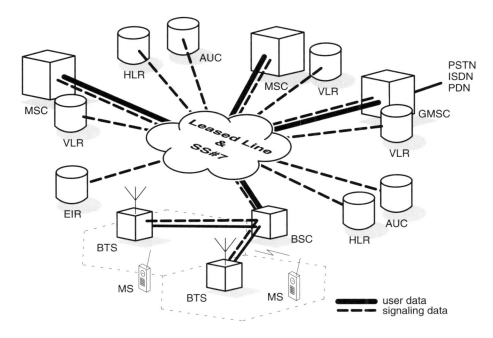

Figure 3.8: Signaling and user data transport in a GSM PLMN

This results in a large number of communication relationships for user data transport and signaling; for simpler structuring and standardization, these relationships have been separated by introducing a number of interfaces (Figure 3.9).

The A interface between BSS and MSC is used for the transfer of data for BSS management, for connection control, and for mobility management. Within the BSS, the Abis interface between BTS and BSC and the air interface Um have been defined.

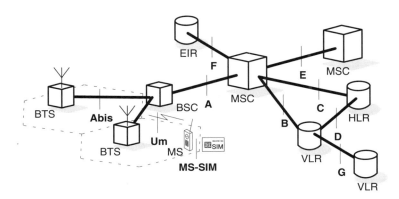

Figure 3.9: Interfaces in a GSM PLMN

This updating of the HLR occurs through the D interface. The D interface is used for the exchange of location-dependent subscriber data and for subscriber management. The VLR informs the HLR about the current location of the mobile subscriber and reports the current MSRN. The HLR transfers all the subscriber data to the VLR that is needed to give the subscriber his or her usual customized service access. The HLR is also responsible for giving a cancellation request for the subscriber data to the old VLR once the acknowledgment for the location update arrives from the new VLR. If, during location updating, the new VLR needs data from the old VLR,it is directly requested over the G interface. Furthermore, the identity of subscriber or equipment can be verified during a location update; for requesting and checking the equipment identity, the MSC has an interface F to the EIR.

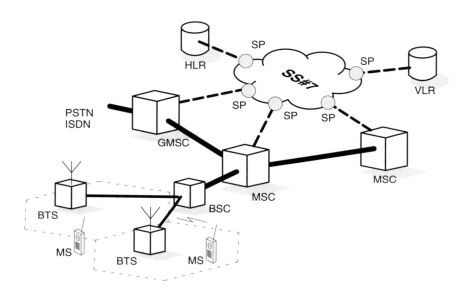

Figure 3.10: Basic configuration of a GSM PLMN

An MSC has two more interfaces besides the A and B interfaces, namely the C and E interfaces. Charging information can be sent over the C interface to the HLR. Besides this, the MSC must be able to request routing information from the HLR during call setup, for calls from the mobile network as well as for calls from the fixed network. In the case of a call from the fixed network, if the fixed network's switch cannot interrogate the HLR directly, initially it routes the call to a gateway MSC (GMSC), which then interrogates the HLR. If the mobile subscriber changes during a conversation from one MSC area to another, a handover needs to be performed between these two MSCs, which occurs across the E interface.

As was already mentioned, the configuration of a PLMN is largely left to the network operator. Figure 3.10 shows a basic configuration of a GSM mobile communication network. This basis configuration contains a central HLR and a central VLR. All data base transactions (updates, inquiries, etc.) and handover transactions between the MSC are performed with the help of the *Mobile Application Part* (MAP) over the SS#7 network. For this purpose, each MSC and register is known as a *Signaling Point* (SP) and is known by its *Signaling Point Code* (SPC) within the SS#7 network.

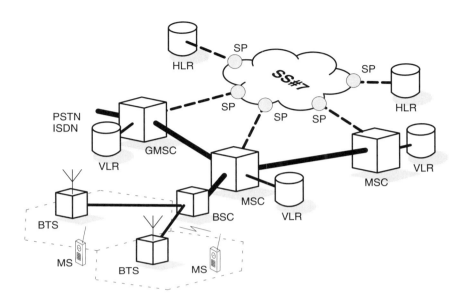

Figure 3.11: Configuration of a GSM PLMN with a VLR for each MSC

The VLR is mainly a database which stores the location information of the mobile stations. At each change of the location area, this information must be updated. Furthermore, this database has to be interrogated: the MSC needs subscriber parameters besides location data for successful connection setup, such as service restrictions and supplementary services to be activated. Thus there is a significant message traffic between MSC and VLR, which constitutes an ensuing load on the signaling network. It is therefore logical, that these two functional units are combined in one physical unit, i.e. to realize the VLR in distributed form and to associate a VLR with each MSC (Figure 3.11). The message traffic between MSC and VLR then need not be transported through the SS#7 network.

One could go one step further and also distribute the database of the HLR and thus introduce several HLRs in a mobile network. This is especially interesting for a growing pool of subscribers, since a centralized database leads to a high traffic load for this database. If there are several HLR in a PLMN, the network operator has to define an association rule between MSISDN and HLR, such that for incoming calls the routing information to an MSISDN can be derived from the associated HLR. One possible association is geographic partitioning of the whole subscriber identification space (SN field in the MSISDN, see Section 3.2.3), where for example, the first two digits of the SN indicate the region and the associated HLR. In extreme cases, the HLR can be realized with the VLR in a single physical unit. In this case, an HLR would also be associated with each MSC.

4 Services

The services offered at the *User–Network Interface* (UNI) of GSM are patterned after the services offered by the *Integrated Services Digital Network* (ISDN) [22] for fixed terminals tied to telephone lines. GSM services are therefore divided just like ISDN services into three categories: bearer services, teleservices, and supplementary services. A bearer service offers the basic technical capability for the transmission of binary data; i.e. it offers the data transfer between end terminals at reference points R or S of the Reference Model (Figure 4.1, Figure 9.1). Such bearer services are made use of by the teleservices for the transfer of data with higher-level protocols.

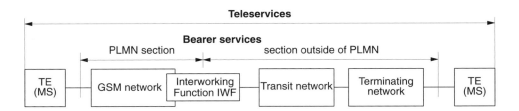

Figure 4.1: Bearer and teleservices

Notice that bearer and teleservices both require special measures not only at the air interface, but also inside the PLMN, which must offer a matching fixed-net infrastructure and special *Interworking Functions* (IWFs). Especially at the level of bearer services, the IWF must provide a mapping of GSM PLMN services within their respective service characteristics onto corresponding bearer services and characteristics of the other networks, such as PSTN and ISDN. Teleservices are end-to-end services, for which there is usually no translation in the IWF. But they do use bearer services, which again need IWF functions.

Bearer and teleservices are carried under the umbrella term *telecommunication services*. The simultaneous use of two telecommunication services is precluded, except for the case of short message services (SMS), which can at least be received during the use of another telecommunication service. Supplementary services are complementing the control and modification of extended services and are only usable in connection with a telecommunication service.

For telecommunication services, the GSM standard includes agreements of target times for their market introduction. This fact is especially important, since GSM is an international standard which aims at worldwide compatibility of mobile stations and networks. Accordingly, only a minimum of services has been defined, which must be offered by the operators at various time phases. For this purpose, the services are divided into the categories *essential* (E) and *additional* (A). Group E must be implemented at the given date by all network operators, whereas the decision about the time of introduction of Group A services is left to the operators. Table 4.1 gives a rough overview over the implementation and introduction phases. The most important services to be implemented and introduced with their respective introduction dates will be described briefly in the following.

Table 4.1: Phases of implementation and introduction

Class	Introduction	Services
E1	1991	Basic operation consisting of telephone services and some appropriate supplementary services
E2	1994	Extended operation with telephone services , first non-speech services (e.g. BS26) and an extended range of supplementary services
E3	1996	Enhanced service range with even more telecommunication and supplementary services

4.1   Bearer Services

The basic services of a GSM network are the foundation for data transmission, i.e. a basic service provides the fundamental technical facilities at the end terminal interface (reference point R) to transport user payloads. The basic services are called transport services [22] or bearer services in ISDN − and therefore in GSM, too. The GSM bearer services offer asynchronous or synchronous data transport capabilities with circuit-switched or packet-switched data rates of 300 to 9600 bit/s, with a 13 kbit/s bearer service for voice.

Bearer services carry only the coding- and application-independent information transport between the user–network interfaces (Figure 4.1); they represent the services of layers 1, 2 and 3 of the OSI Reference Model. The operators of the terminal equipment (TE) can use these services and employ arbitrary higher-level protocols, but they are responsible for the compatibility of the protocols used in the terminal equipment, quite in contrast to teleservices, where the protocols in the terminal equipment are also standardized [22]. An overview of the most important bearer services is given in Table 4.2. Each bearer service has its own number, e.g. BS26 is the bearer service for circuit-switched asynchronous data transfer at 9600 bit/s.

Besides the asynchronous and synchronous circuit-switched data services (BS21 to BS34), packet-switched data services are also provided. These packet services are realized either as asynchronous access to a *Packet Assembler/Disassembler* (PAD), BS41 to BS46, or as direct synchronous *Packet Access*, BS51 to BS53.

The bearer services for GSM data transfer are offered in two fundamentally different modes (Table 4.2): transparent (T) and nontransparent (NT). In the transparent mode, there is a circuit-switched connection between the mobile terminal (TE) and the interworking module in the MSC, from where the connection to other networks is handled. This connection is protected by *Forward Error Correction* (FEC). The most important common characteristics for all transparent services are constant bit rate, constant transport delay, and residual bit error ratio dependent on the current channel conditions. The nontransparent mode activates a special Layer 2 protocol for the additional protection of the data transfer, the *Radio Link Protocol* (RLP) which is specially adapted to the GSM radio channel. This protocol terminates in the mobile station and in the MSC. It uses ARQ procedures to request retransmission of blocks with residual errors which could not be corrected by forward error correction.

Table 4.2: GSM bearer services (excerpt)

Service	Structure	BS no.	Bit rate [bit/s]	Mode	Transmission
Data	asynch	21	300	T or NT	UDI or 3.1 kHz
		22	1200	T or NT	UDI or 3.1 kHz
		23	1200/75	T or NT	UDI or 3.1 kHz
		24	2400	T or NT	UDI or 3.1 kHz
		25	4800	T or NT	UDI or 3.1 kHz
		26	9600	T or NT	UDI or 3.1 kHz
Data	synch	31	1200	T	UDI or 3.1 kHz
		32	2400	T or NT	UDI or 3.1 kHz
		33	4800	T or NT	UDI or 3.1 kHz
		34	9600	T or NT	UDI or 3.1 kHz
PAD	asynch	41	300	T or NT	UDI
		42	1200	T or NT	UDI
		43	1200/75	T or NT	UDI
		44	2400	T or NT	UDI
		45	4800	T or NT	UDI
		46	9600	T or NT	UDI
Packet	synch	51	2400	NT	UDI
		52	4800	NT	UDI
		53	9600	NT	UDI
Altern. speech/ data		61	13 000 or 9600		
Speech followed by data		81	13 000 or 9600		

T, NT = transparent, non transparent   UDI = Unrestricted Digital Information   PAD = Packet Assembler/Disassembler
asynch, synch = asynchronous, synchronous

This gives a much more significant reduction in the residual error rate. In essence an error-free information transport is achieved, hence it is approximately independent of the momentary channel conditions. However, with changing error behavior of the radio channel, the frequency of block repetitions also varies, and thus the average transfer delay and the net bit rate of the data service vary too. Activation of the nontransparent data service is especially interesting for rapidly moving mobile stations or for cases of bad radio conditions, where high fading rates and deep fading/shadow holes occur. In such situations, a meaningful transport of user data with transparent mode can become impossible. At the expense of a net data rate decrease, the nontransparent mode then still allows a reliable data transport.

The GSM bearer services 21 to 53 are further categorized into *Unrestricted Digital Information* (UDI) and 3.1 kHz (Table 4.2). The services differ mainly in the way in which they are handled outside of the PLMN, i.e. the kind of interworking function that needs to be activated. The UDI service category corresponds to the *Unrestricted Digital Information* (UDI) of ISDN and supplies a channel for the unrestricted transfer of digital information. The data transfer is unrestricted in the sense that no bit patterns are reserved or explicitly excluded from transmission. The 3.1 kHz category is used to activate in the MSC an interworking function for 3.1 kHz audio and to select a modem. Within the GSM PLMN (from user-network access to the interworking function), the data are still transferred as *Unrestricted Digital Information.* The designation "3.1 kHz" rather refers to the fact that the transfer outside of the PLMN uses a service "3.1 kHz Audio". This service is offered by conventional PSTN as well as by ISDN networks. For transfer with this service, the data has to be converted in the IWF of the MSC with a modem to an audio signal with a bandwidth of 3.1 kHz.

Further important GSM bearer services contain voice (telephone) service (BS61 to BS81), which can be (multiple times) changed during a call at the request of the user to a data service (alternate speech/data).An other alternative is that the user at first establishes a voice connection and then changes to a data connection, which cannot be changed back to voice (speech followed by data).

4.2 Teleservices

On top of the bearer services, which can be used by themselves, a number of teleservices have been defined. The most important categories are (Table 4.3) speech, *Short Message Service* (SMS), access to *Message Handling Systems* (MHSs) and to videotext, teletext and facsimile transfer.

Voice services had to be implemented by each operator in the start-up phase (E1) by 1991. In this category, two teleservices were distinguished: regular telephone service (TS11) and emergency service (TS12). For transmission of the digitally coded speech signals, both services use a bidirectional, symmetric, full-duplex point-to-point connection, which is set up on user demand. The sole difference between TS11 and TS12 teleservices is that regular service requires an international IWF, whereas the emergency service stays within the boundaries of a national network.

As teleservice for the second implementation phase (E2), implementation of transparent fax service (TS61) for Group 3 fax was planned. The fax service is called *transparent* because it uses a transparent bearer service for the transmission of fax data. The coding and transmission of the facsimile data uses the fax protocol according to the ITU-T recommendation T30. The network operator also has the option to implement TS61 on a nontransparent bearer service in order to improve the transmission quality. TS61 is transmitted over a traffic channel that is alternately used for voice or fax. Another optional alternative is designated as *fax transfer with automatic call acceptance* (TS61). This service can be offered by a network operator when *multinumbering* is used as the interworking solution. In the case of multinumbering, a subscriber is assigned several MSISDN numbers, and a separate interworking profile is stored for each of them. In this way a specific teleservice can be associated with each MSISDN, the fax service being one of them. If a mobile subscriber is called on his "GSM-fax number", the required resources in the IWF of the MSC as well as in the MS can be activated; whereas in the case of TS61, fax calls arrive with the same number as voice calls (no multinumbering) and have to be switched over to fax reception manually.

Table 4.3: GSM teleservices (excerpt)

Category	TS no.	Service		Class
Speech	11	Telephone		E1
	12	Emergency call		E1
Short Message Services	21	Short Message Mobile Terminated, Point to Point		E3
	22	Short Message Mobile Originated, Point to Point		A
	23	Short Message Cell Broadcast		–
MHS access	31	Access to Message Handling Systems		A
Videotex access	41	Videotex access profile 1		A
	42	Videotex access profile 2		A
	43	Videotex access profile 3		A
Teletext transmission	51	Teletext		A
Fax transmission	61	Speech and fax group 3	T	E2
		alternating	NT	A
	62	Fax group 3 automatic	T	–
			NT	–

Another teleservice which was assigned high priority in the service implementation strategy is the capability to receive or send short messages at the mobile station: *Short Message Service* (SMS), TS21 and TS22. This service was supposed to be offered in the third phase (E3) at the latest from 1996 on all GSM networks. TS21 is the point-to-point version of the SMS, which allows a single station to be sent a message of up to 160 characters. Conversely, TS22 has been defined as an optional implementation of the capability to send short messages from a mobile station. The combinations of SMS with other added-value services, e.g. mailbox systems with automatic notification of newly arrived messages or the transmission by short message of incurred charges, clearly show how the services offered by GSM networks go significantly beyond the services offered in fixed networks.

For SMS, the network operator has to establish a service center which accepts short messages from the fixed network and processes them in a store-and-forward mode. The interface has not been specified and can be by DTMF signaling, special order, email, fax, etc. The delivery can be time-shifted and is of course independent of the current location of the mobile station. Conversely, a service center can accept short messages from mobile stations which can also be forwarded to subscribers in the fixed network, for example by fax or email. The transmission of short messages uses a connectionless, protected, packet-switching protocol. The reception of a message must be acknowledged by the mobile station or the service center; in case of failure, retransmission occurs.

TS21 and TS22 are the only teleservices which can be used simultaneously with other services, i.e. short messages can also be received or transmitted during an ongoing call.

A further variation of the SMS is the *Cell Broadcast Service* TS23, *Short Message Service Cell Broadcast* (SMSCB). SMSCB messages are broadcast only in a limited region of the network. They can only be received by mobile stations in *idle mode,* and reception is not acknowledged. A mobile station itself can not send SMSCB messages. With this service, messages contain a category designation, so that mobile stations can select categories of interest which they want to receive and store. The maximum length of SMSCB messages is 93 characters, but by using a special reassembly mechanism, the network can transmit longer messages of up to 15 subsequent SMSCB messages.

Further teleservices such as access to a message handling system according to ITU-T recommendation X.400 or access to videotext systems are provided as options only. In 1996, due to lack of demand, there were hardly any network operator implementations of these services. Instead, the extension of asynchronous data services (bearer services) has been forcefully promoted, such that it is currently possible for mobile computers to perform worldwide fax transmissions and data communication.

4.3 Supplementary Services

The supplementary services in GSM correspond to the supplementary services of ISDN with regard to service and performance characteristics. They can be used only in connection with a teleservice, i.e. they modify or supplement the functionality of a GSM telecommunication service (bearer or teleservice). Besides the improved network organization, the introduction of numerous ISDN-like supplementary services is the main feature of GSM Phase 2. Some GSM supplementary services are identical or similar to those offered in ISDN, but their implementation is often much more complex due to the added mobility. Beyond that, GSM offers new service characteristics which are available in ISDN networks only in restricted form or not at all.

For Phase 1 of GSM, only a small set of supplementary services concerning call forwarding and call restriction was defined (Table 4.4). If a mobile station activates call forwarding, then calls are not switched through to this MS, but forwarded to a configurable extension. Several variations can be distinguished: first, unconditional call forwarding (CFU) where all calls are diverted; then conditional call forwarding when calls are only forwarded under special conditions, such as when the MS is busy (CFB) or is not reachable (CFNRc), possibly because it is powered off or outside any covered network area.

The network operators usually offer a voice mailbox service in connection with call for-warding. This consists of an answering machine function within the network, which offers recording of voice messages for later retrieval by the subscriber for incoming calls, if the call forwarding feature has been activated. This kind of service offering clearly goes be-yond what fixed ISDN networks are offering. Of course, call forwarding can also be di-rected at another target than the voice mailbox.

GSM Phase 1 also introduced supplementary services for barring of either outgoing or in-coming calls. In this case there are also several variants. For example, all calls can be barred outgoing (BAOC) or barred incoming (BAIC), or it may be only outgoing international calls which are barred (BOIC), or perhaps incoming calls that might cause charges such as calls to an MS which is roaming outside its home network (BIC-Roam)

Table 4.4: Overview of GSM supplementary services (GSM Phase 1)

Category	Abbrev.	Service	Class
Call Offering	CFU	Call Forwarding Unconditional	E1
	CFB	Call Forwarding on Mobile Subscriber Busy	E1
	CFNRy	Call Forwarding on No Reply	E1
	CFNRc	Call Forwarding on Mobile Subscriber Not Reachable	E1
Call Restriction	BAOC	Barring of All Outgoing Calls	E1
	BOIC	Barring of Outgoing International Calls	E1
	BAIC	Barring of All Incoming Calls	E1
	BOIC-exHC	Barring of Outgoing International Calls Except Calls to Home PLMN	A
	BIC-Roam	Barring of Incoming Calls when Roaming Outside the Home PLMN	A

In the course of further evolution of the GSM standard, the menu of services known from ISDN is being made available in stages [22] and supplemented by some new GSM-specific performance characteristics. In Phase 2, which was standardized in 1996 and is under de-ployment, there are already some new services (Table 4.5), such as call waiting (CW) or hold (HOLD), which enable performing brokerage functions.

Two very powerful supplementary services are conference calling (CONF) allowing the in-terconnection of several subscribers in one call, and call transfer (CT) which allows a call to be passed to a third party. Of special interest in connection with call waiting and call transfer services are the supplementary services of the number identification category (Table 4.5). The calling line identification presentation (CLIP) lets the calling party's MSISDN number appear on the display of the called party, but the calling party can pre-vent this by activating the supplementary service calling line identification restriction (CLIR), in case the caller does not want to disclose his or her number. The eventually reached number may not always be the number called by the calling party, e.g. in the case of a call transfer. With the supplementary service connected line identification presenta-tion (COLP) the caller can request to be shown the reached extension, but the called party can prevent this announcement by using the connected line identification restriction (COLR). The inquiry of current charges is also offered with a supplementary service, as well as reverse charging (REVC), which allows the called party to assume the charges for the call. These features are clearly responsible for providing a lot more calling comfort in GSM networks than ISDN networks are offering, even though digital technology enables all of them in both.

Table 4.5: Overview of GSM supplementary services (GSM Phase 2)

Category	Abbrev.	Service	Class
Number Identification	CLIP	Calling Line Identification Presentation	A
	CLIR	Calling Line Identification Restriction	A
	COLP	Connected Line Identification Presentation	A
	COLR	Connected Line Identification Restriction	A
	MCI	Malicious Call Identification	A
Call Offering	CT	Call Transfer	A
	MAH	Mobile Access Hunting	A
Community of Interest	CUG	Closed User Group	A
Charging	AoC	Advice of Charge	E2
	FPH	Freephone Service	A
	REVC	Reverse Charging	A
Add. Information Transfer	UUS	User-to-User Signaling	A
Call Completion	CW	Call Waiting	E3
	HOLD	Call Hold	E2
	CCBS	Completion of Call to Busy Subscriber	A
Multi-Party	3PTY	Three-Party Service	E2
	CONF	Conference Calling	E3

The standardization and further development of GSM systems, however, is not completed with Phase 2 and continues to proceed. This process is generally known under the name *GSM Phase 2+* (see Section 11.2). There is special focus on bearer services with higher bit rates up to 64 kbit/s. Also connectionless packet-switched data communication at the air interface is being investigated and standardized to provide a *General Packet Radio Service* (GPRS). Both are especially interesting for diverse applications when the mobile data communication is typically characterized by bursty traffic, so a complete traffic channel is not required for the duration of a connection.

5 Air Interface – Physical Layer

In GSM, the physical layer which resides on the first of the seven layers of the OSI Reference Model, contains very complex functions. On top of the physical channels, a series of logical channels have been defined at the *User–Network Interface* (UNI) to perform a multiplicity of functions, e.g. signaling, broadcast of general system information, synchronization, channel assignment, paging, payload transport. In order to achieve the best bandwidth efficiency, the logical control channels are mapped onto physical channels in certain time-multiplexed combinations. This means that they have to share a physical channel, whereas user payload data is allocated a dedicated full-rate or half-rate channel. The logical channels are explained in Section 5.1, which serves as a foundation for understanding the signaling procedures at the air interface. Not all logical channels may be used simultaneously on a physical channel, but only a certain combination of them. The realization of the physical channels and the mapping of logical onto physical channels follows in Sections 5.2 to 5.4, where the higher-level multiplexing of logical channels into multiframes, etc., is also covered. The conclusion of the section contains a discussion of the most important control mechanisms for the air interface and a power-up scenario with the sequence of events occurring, from when a mobile station is turned on to when it is in a synchronized state ready to transmit.

Table 5.1: Classification of logical channels in GSM

Traffic Channels	Signaling Channels		
Bidirectional	**Unidirectional, Downlink**	**Unidirectional, Down- or Uplink**	**Bidirectional**
Traffic Channel TCH	**Broadcast Channel BCH**	**Common Control Channel CCCH**	**Dedicated Control Channel DCCH**
Full-rate channel Bm	Broadcast Control Channel BCCH	Random Access Channel RACH	Standalone Dedicated Control Channel SDCCH
Half-rate channel Lm	Synchronization Channel SCH	Access Grant Channel AGCH	**Associated Control Channel ACCH**
	Frequency Correction Channel FCH	Paging Channel PCH	Slow Associated Control Channel SACCH
			Fast Associated Control Channel FACCH

5.1   Logical Channels

On Layer 1 of the OSI Reference Model, GSM defines a series of logical channels, which are made available either in an unassigned random access mode or in a dedicated mode assigned to a specific user. Logical channels are divided into two categories (Table 5.1).

The first category comprises the traffic channels:

- *Traffic Channel* (TCH)
  The traffic channels are used for the transmission of user payload data (speech, fax, data). They do not carry any control information of Layer 3. Communication over a TCH can be circuit-switched or packet-switched. In the circuit-switched case, the TCH provides a transparent data connection or a connection that is specially treated according to the carried service (e.g. telephony). For the packet-switched mode, the TCH carries user data of OSI Layers 2 and 3 according to the recommendations of the X.25 standard or similar standard packet protocols. A TCH may either be fully used (full-rate TCH) or be split into two half-rate channels (half-rate TCH), which can be allocated to different subscribers.

Following ISDN terminology, the GSM traffic channels are also designated as Bm channel (mobile B channel) or Lm channel (Lower-rate mobile channel, with half the bit rate). A Bm channel is a TCH for the transmission of bit streams of either 13 kbit/s of digitally coded speech or of data streams at 12, 6, or 3.6 kbit/s. Lm channels are TCH channels with less transmission bandwidth than Bm channels and transport speech signals of half the bit rate (half-rate TCH) or bit streams for data services with 6 or 3.6 kbit/s.

The other category comprises the signaling channels, which are also called Dm channels (mobile D channel). They are further divided into

- *Broadcast Channel* (BCH)
  See below.

- *Common Control Channel* (CCCH)
  A CCH is a unidirectional point-to-multipoint signaling channel to deal with access management functions. This includes the assignment of dedicated channels and paging to exactly localize a mobile station.

- *Dedicated/Associated Control Channel* (DCCH/ACCH)
  A dedicated control channel (DCCH) is a bidirectional point-to-point signaling channel. An associated control channel is also a dedicated control channel, but it is assigned only in connection with a traffic channel TCH or a standalone dedicated control channel (SDCCH).

The category *Broadcast Channel* (BCH) consists of three channels:

- *Broadcast Control Channel* (BCCH)
  The BCCH is a unidirectional point-to-multipoint channel between *Base Station Subsystem* (BSS) and *Mobile Station* (MS). On this channel, a series of information elements is broadcast to the mobile stations which characterize the organization of the radio network, such as radio channel configurations (of the currently used cell as well as of the neighboring cells), synchronization information (frequencies as well as frame numbering), registration identifiers (LAI, CI, BSIC). In particular,

this includes information about the structural organization (formats) of the CCCH of the local base station. The BCCH is broadcast on the first frequency assigned to the cell (BCCH carrier).

- *Frequency Correction Channel* (FCCH)
  See below (Section 5.2.2, frequency correction burst).

- *Synchronization Channel* (SCH)
  The SCH broadcasts information to identify a BTS, i.e. *Base Station Identity Code* (BSIC); see 3.2.9. The SCH also broadcasts data for the frame synchronization of an MS , i.e. *Reduced Frame Number* (RFN) of the TDMA frame; see Section 5.3.1.

SCH and FCCH are only visible within protocol Layer 1, since they are only needed for the operation of the radio subsystem. There is no access to them from Layer 2. In spite of this fact, the SCH messages contain data which are needed by Layer 3 for the administration of radio resources. These two channels are always broadcast together with the BCCH.

The *Common Control Channel* comprises the following:

- *Random Access Channel* (RACH)
  The RACH is the uplink portion of the CCCH. It is accessed from the mobile stations in a cell without reservation in a competitive multi-access mode using the principle of slotted Aloha [27], to ask for a dedicated signaling channel (SDCCH) for exclusive use by one mobile station for one signaling transaction.

- *Access Grant Channel* (AGCH)
  The AGCH is the downlink part of the CCCH. It is used to assign an SDCCH or a TCH to a mobile station.

- *Paging Channel* (PCH)
  The PCH is also part of the downlink of the CCCH. It is used for paging to find specific mobile stations.

The group of *Dedicated/Associated Control Channels* (D/ACCH) comprises the following:

- *Standalone Dedicated Control Channel* (SDCCH)
  The SDCCH is a dedicated point-to-point signaling channel (DCCH) which is not tied to the existence of a TCH, i.e. standalone. The SDCCH is requested from the MS via the RACH and assigned via the AGCH. After the completion of the signaling transaction, the SDCCH is released and can be reassigned to another MS. Examples of signaling transactions which use an SDCCH are the updating of location information or parts of the connection setup until the connection is switched through (see Figure 5.1).

- *Slow Associated Control Channel* (SACCH)
  An SACCH is always assigned and used with a TCH or an SDCCH. The SACCH carries information for the optimal radio operation, e.g. commands for synchronization and transmitter power control and reports on channel measurements (Section 5.5). Data must be transmitted continuously over the SACCH since the arrival of SACCH packets is taken as proof of the existence of the physical radio connection (Section 5.5.3). When there is no signaling data to transmit, the MS sends a *measurement report* with the current results of the continuously conducted radio signal level measurements (Section 5.5.1).

- *Fast Associated Control Channel* (FACCH)
  By using dynamic preemptive multiplexing on a TCH, additional bandwidth can be made available for signaling. The signaling channel created this way is called FACCH. It is only assigned in connection with a TCH, and its short-time usage goes at the expense of the user data transport.

In addition to these channels, a *Cell Broadcast Channel* (CBCH) is defined, which is used to broadcast the messages of the *Short Message Service Cell Broadcast* (SMSCB). The CBCH shares a physical channel together with the SDCCH.

Table 5.2 gives an overview over the logical channels of Layer 1, the available bit rates, block lengths used, and the intervals between transmission of blocks. Notice that the logical channels can suffer from substantial transmission delays depending on the respective use of forward error correction (channel coding and interleaving, see Section 6.2 and Table 6.8).

Table 5.2: Logical channels of GSM Protocol Layer 1

Channel type	Net data throughput [kbit/s]	Block length [bit]	Block distance [ms]
TCH (full-rate speech)	13.0	182+78	20
TCH (half-rate speech)	standardi-zation in progress	standardi zation in progress	standardi-zation in progress
TCH (data, 9.6 kbit/s)	12.0	60	5
TCH (data, 4.8 kbit/s)	6.0	60	10
TCH (data, $\leq$ 2.4 kbit/s)	3.6	72	20
Full-rate FACCH	9.2	184	20
Half-rate FACCH	4.6	184	40
SDCCH	598/765	184	3060/13
SACCH (with TCH)	115/300	168+16	480
SACCH (with SDCCH)	299/765	168+16	6120/13
BCCH	598/765	184	3060/13
AGCH	n*598/765	184	3060/13
PCH	p*598/765	184	3060/13
RACH	r*27/765	8	3060/13

The (logical) channels are not simultaneously usable at the radio interface. They can only be deployed in certain combinations and on certain physical channels. GSM has defined several channel configurations (Table 5.3), which are realized and offered by the base stations. Depending on its current state, a mobile station can only use a subset of the logical channels in Table 5.3 which are offered at the air interface by the base station. Here too, there are seven channel configurations defined which can be used by a mobile station; however, a mobile station uses the channels offered by a base station only in the combinations indicated in Table 5.4.

Table 5.3: Channel combinations offered by the base station

	Bm	Lm	Lm	FACCH	SACCH	SDCCH	FCCH	SCH	BCCH	CCCH
B1									▓	▓
B2							▓	▓	▓	▓
B3					▓	▓	▓	▓	▓	▓
B4					▓	▓				
B5	▓			▓	▓					
B6		▓		▓	▓					
B7		▓	▓	▓	▓					

The combination M1 from Table 5.4 is only used in the phase when no physical connection exists, i.e. immediately after the power-up of the mobile station or after a disruption due to unsatisfactory radio signal conditions. Channel combinations M2 and M3 are used by active mobile stations at rest. In phases requiring a dedicated signaling channel, a mobile station uses the combination M4, whereas M5 to M7 are used when there is a traffic channel up.

Table 5.4: Channel combinations used by the mobile station

	Bm	Lm	Lm	FACCH	SACCH	SDCCH	BCCH	CCCH
M1							▓	
M2								▓
M3						▓		
M4					▓	▓		
M5	▓			▓	▓			
M6		▓		▓	▓			
M7		▓	▓	▓	▓			

It is also important to notice that, in each channel combination (M1, M2, M3), there is always an SACCH allocated, either with a TCH or with an SDCCH, which accounts for the attribute "associated".

Figure 5.1 shows in an example for an incoming call connection setup at the air interface how the various logical channels are used in principle. The mobile station is called via the PCH and requests a signaling channel on the RACH. It gets the SDCCH through an IMMEDIATE ASSIGNMENT message on the AGCH. Then follow authentication, start of ciphering, and start of setup over the SDCCH. An ASSIGNMENT COMMAND message gives the traffic channel to the mobile station, which acknowledges its receipt on the FACCH of this traffic channel. The FACCH is also used to continue the connection setup.

MS   BSS/MSC

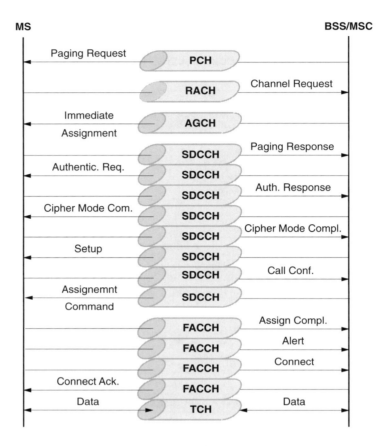

Figure 5.1: Logical channels and signaling (connection setup for an incoming call)

5.2 Physical Channels

5.2.1 Modulation

The modulation on the radio channel is known by the abbreviation GMSK, which stands for *Gaussian Minimum Shift Keying*. It belongs to a family of continuous-phase modulation procedures, which have the special advantages of a narrow transmitter power spectrum with low adjacent channel interference on one hand and a constant amplitude envelope on the other hand, which allows use of simple amplifiers in the transmitters without special linearity requirements (class C amplifiers). Such amplifiers are especially inexpensive to manufacture, have high degree of efficiency, and therefore allow longer operation on a battery charge [30][32].

The digital modulation procedure for the GSM air interface comprises several steps for the generation of a high-frequency signal from channel-coded and enciphered data blocks (Figure 5.2).

Figure 5.2: Steps of GSM digital modulation

The data d_i arrives at the modulator with a bit rate of $1625/6$ kbit/s $= 270.83$ kbit/s (gross data rate) and are first differential-coded:

$$\hat{d}_i = (d_i + d_{i-1}) \bmod 2; \quad d_i \in (0; 1)$$

From this differential data, the modulation data is formed, which represents a sequence of Dirac pulses:

$$a_i = 1 - 2\hat{d}_i$$

This bipolar sequence of modulation data is fed into the transmitter filter − also called a frequency filter − to generate the phase $\varphi(t)$ of the modulation signal.

The impulse response $g(t)$ of this linear filter is defined by the convolution of the impulse response $h(t)$ of a Gaussian low-pass with a rectangular step function:

$$g(t) = h(t) * \mathrm{rect}\,(t/T)$$

$$\mathrm{rect}\,(t/T) = \begin{cases} 1/T & \text{for} \quad |t| < T/2 \\ 0 & \text{for} \quad |t| \ge T/2 \end{cases}$$

$$h(t) = \frac{1}{\sqrt{2\pi}\,\sigma T} \exp\!\left(\frac{-t^2}{2\sigma^2 T^2}\right) \qquad\qquad \sigma = \frac{\sqrt{\ln 2}}{2\pi BT}; \; BT = 0,3$$

In the equations above, B is the 3dB bandwidth of the filter $h(t)$ and T the bit duration of the incoming bit stream. The employed rectangular step function and the impulse response of the Gaussian lowpass are shown in Figure 5.3, and the resulting impulse response $g(t)$ of the transmitter filter is given in Figure 5.4 for some values of BT. Notice that with decreasing BT the impulse response becomes broader. For $BT \to \infty$ it converges to the rect() function.

Figure 5.3: Impulse responses for the building blocks of the GMSK transmitter filter

In essence, this modulation consists of a *Minimum Shift Keying* (MSK) procedure, where the data is filtered through an additional Gaussian lowpass before *Continuous Phase Modulation* (CPM) with the rectangular filter [32]. Accordingly it is called *Gaussian MSK* (GMSK). The Gaussian lowpass filtering has the effect of additional smoothing, but also of broadening the impulse response $g(t)$. This means that, on one hand the power spectrum of the signal is made narrower, but on the other hand the individual impulse responses are "smeared" across several bit durations, which leads to increased intersymbol interference. This partial-response behavior has to be compensated for in the receiver by means of an equalizer [32].

The phase of the modulation signal is the convolution of the impulse response $g(t)$ of the frequency filter with the Dirac impulse sequence a_i of the stream of modulation data:

$$\varphi(t) \;=\; \sum_i a_i \,\pi\eta \int_{-\infty}^{t-iT} g(u)\,du$$

with the modulation index at $\eta = 1/2$, i.e. the maximal phase shift is $\pi/2$ per bit duration. Accordingly, GSM modulation is designated as 0.3-GMSK with a $\pi/2$ phase shift. The phase $\varphi(t)$ is now fed to a phase modulator. The modulated high-frequency carrier signal can then be represented by the following expression, where E_c is the energy per bit of the modulated data rate, f_0 the carrier frequency, and φ_0 is a random phase component staying constant during a burst:

$$x(t) \;=\; \sqrt{\frac{2E_c}{T}} \,\cos(2\pi f_0 t + \varphi(t) + \varphi_0)$$

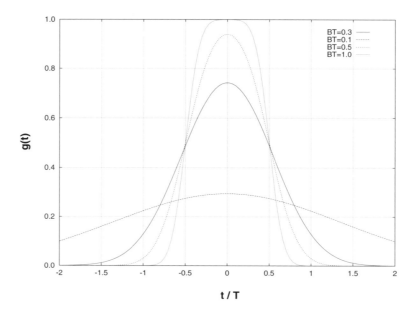

Figure 5.4: Impulse response $g(t)$ of the frequency filter (transmitter filter)

5.2.2   Multiple Access, Duplexing, and Bursts

On the physical layer (OSI Layer 1), GSM uses a combination of FDMA and TDMA for multiple access. Two frequency bands 45 MHz apart have been reserved for GSM operation (Figure 5.5): 890–915 MHz for transmission from the mobile station, i.e. uplink, and 935–960 MHz for transmission from the base station, i.e. downlink. Each of these bands of 25 MHz width is divided into 124 single carrier channels of 200 kHz width. This variant of FDMA is also called *Multi-Carrier* (MC). In each of the uplink / downlink bands there remains a guardband of 200 kHz. Each *Radio Frequency Channel* (RFCH) is uniquely numbered, and a pair of channels with the same number form a duplex channel with a duplex distance of 45 MHz (Figure 5.5).

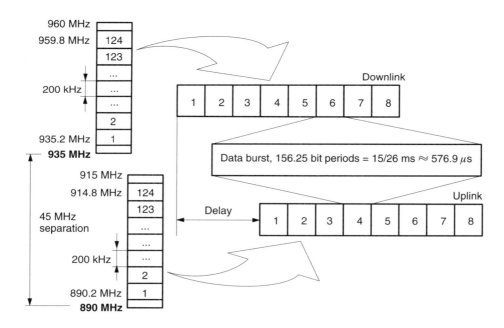

Figure 5.5: Carrier frequencies, duplexing, and TDMA frames

A subset of the frequency channels, the *Cell Allocation* (CA), is allocated to a base station, i.e. to a cell. One of the frequency channels of the CA is used for broadcasting the synchronization data (FCCH and SCH) and the BCCH. Therefore this channel is also called the BCCH Carrier (see Section 5.4). Another subset of the cell allocation is allocated to a mobile station, the *Mobile Allocation* (MA). The mobile allocation is used among others for the optional frequency hopping procedure (Section 5.2.3). Countries or areas which allow more than one mobile network to operate in the same area of the spectrum must have a licensing agency which distributes the available frequency number space (e. g. the Federal Communication Commission in the USA, or the Federal Office of a PTT, or the BAPT in Germany), in order to avoid collisions and to allow the network operators to perform independent network planning. Here is an example for a possible division: Operator A uses RFCH 2–13, 52–81 and 106–120, whereas operator B receives RFCH 15–50, 83–103, in which case RFCH 1, 14, 51, 82, 104, 105, and 121–124 are left unused as additional guard bands.

Each of these 200 kHz channels carries eight TDMA channels by dividing each of them into eight time slots. The eight time slots in these TDMA channels form a TDMA frame (Figure 5.5). The TDMA frames of the uplink are transmitted with a delay of three time slots with regard to the downlink (see Figure 5.7). A mobile station uses the same time slots in the uplink as in the downlink, i.e. the time slots with the same number (TN). Because of the shift of three time slots, the MS does not have to send at the same time as it receives, and therefore does not need a duplex unit. This reduces the high-frequency requirements for the front end of the mobile and allows it to be manufactured as a less expensive and more compact unit.

So besides the separation into uplink and downlink bands, *Frequency Division Duplex* (FDD) with a distance of 45 MHz, the GSM access procedure contains a *Time Division Duplex* (TDD) component. Thus the MS does not need its own high-frequency duplexing unit, which again reduces cost as well as energy consumption.

Each time slot of a TDMA frame lasts for a duration of 156.25 bit times and, if used, contains a data burst. The time slot lasts 15/26 ms = 576.9 μs; so a frame takes 4.613 ms.

The same result is also obtained from the GMSK procedure, which realizes a gross data transmission rate of 270.83 kbit/s per carrier frequency.

There are five kinds of burst (Figure 5.6):

- *Normal Burst* (NB)
  The normal burst is used to transmit information on traffic and control (except RACH) channels. The individual bursts are separated from each other by guard periods during which no bits are transmitted. At the start and end of each burst are three tail bits which are always set to logical "0". These bits fill a short time span during which transmitter power is ramped up or ramped down and during which no data transmission is possible. Furthermore, the initial zero bits are also needed for the demodulation process. The *Stealing Flags* (SF) are signaling bits which indicate whether the burst contains traffic data or signaling data. They are set to allow use of single time slots of the TCH in preemptive multiplexing mode, e.g. when, during a handover, fast transmission of signaling data on the FACCH is needed. This causes a loss of user data, i.e. these time slots are "stolen" from the traffic channel, hence the name "stealing flag". A normal burst contains besides the synchronization and signaling bits (Figure 5.6) two blocks of 57 bits each of error-protected and channel-coded user data separated by a 26-bit midamble. This midamble consists of predefined, known bit patterns, the training sequences, which are used for channel estimation to optimize reception with an equalizer and for synchronization. With the help of these training sequences, the equalizer eliminates or reduces the intersymbol interferences which are caused by propagation time differences of the multipath propagation. Time differences of up to 16 μs can be compensated for. Eight different training sequences are defined for the NB which are designated by the *Training Sequence Code* (TSC). Initially, the TSC is obtained when the *Base Station Color Code* (BCC) is obtained, which is transmitted as part of the BSIC (see Section 3.2.9). Beyond that, training sequences can be individually assigned to mobile stations. In this case the TSC is contained in the Layer 3 message of the channel assignment (TCH or SDCCH). That way the base station tells a mobile station which training sequence it should use with normal bursts of a specific traffic channel.

- *Frequency Correction Burst* (FB)
  This burst is used for the frequency synchronization of a mobile station. The repeated transmission of FBs is also called the *Frequency Correction Channel* (FCCH). Tail bits as well as data bits are all set to 0 in the FB. Due to the GSM modulation procedure (0.3-GMSK) this corresponds to broadcasting an unmodulated carrier with a frequency shift of 1625 / 24 kHz above the nominal carrier frequency. This signal is periodically transmitted by the base station on the BCCH carrier. It allows time synchronization with the TDMA frame of a mobile station as well as the exact tuning to the carrier frequency. Depending on the stability of its own reference clock, the mobile can periodically resynchronize with the base station using the FCCH.

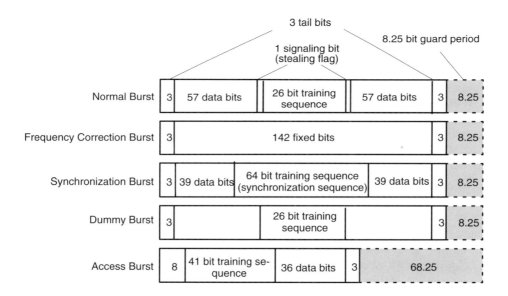

Figure 5.6: Bursts of the GSM TDMA procedure

- *Synchronization Burst* (SB)
  This burst is used to transmit information which allows the mobile station to synchronize time-wise with the base station BTS. Besides a long midamble, this burst contains the running number of the TDMA frame, the *Reduced TDMA Frame Number* (RFN) and the BSIC; the RFN is covered in Section 5.3. Repeated broadcasting of synchronization bursts is considered as the *Synchronization Channel* (SCH).

- *Dummy Burst* (DB)
  This burst is transmitted on one frequency of the cell allocation CA, when no other bursts are to be transmitted. The frequency channel used is the same one that carries the BCCH, i.e. it is the BCCH carrier. This ensures that the BCCH transmits a burst in each time slot which enables the mobile station to perform signal power measurements of the BCCH, a procedure also known as *quality monitoring*.

- *Access Burst* (AB)
  This burst is used for random access to the RACH without reservation. It has a guard period significantly longer than the other bursts. This reduces the probability of collisions, since the mobile stations competing for the RACH are not (yet) time-synchronized.

A single user gets one-eighth or 33.9 kbit/s of the gross data rate of 270.83 kbit/s. Considering a normal burst, 9.2 kbit/s are used for signaling and synchronization, i.e. tail bits, stealing flags and training sequences, including guard periods. The remaining 24.7 kbit/s are available for the transmission of (raw) user or control data on the physical layer.

5.2.3 Optional Frequency Hopping

Mobile radio channels suffer from frequency-selective interferences, e.g. frequency-selective fading due to multipath propagation phenomena. This selective frequency interference can increase with the distance from the base station, especially at the cell boundaries and under unfavorable conditions. Frequency hopping procedures change the transmission frequencies periodically and thus average the interference over the frequencies in one cell. This leads to a further improvement of the *Signal-to-Noise Ratio* (SNR) to a high enough level for good speech quality, so that conversations with acceptable quality can be conducted. GSM systems achieve a good speech quality with an SNR of about 11 dB. With frequency hopping a value of 9 dB is sufficient.

GSM provides for an optional frequency hopping procedure which changes to a different frequency with each burst; this is known as *slow frequency hopping*. The resulting hopping rate is about 217 changes per second, corresponding to the TDMA frame duration. The frequencies available for hopping, the *hopping assignment*, are taken from the cell allocation. The principle is illustrated in Figure 5.7, showing the time slot allocations for a full-rate TCH. The exact synchronization is determined by several parameters: the *Mobile Allocation* (MA), a *Mobile Allocation Index Offset* (MAIO), a *Hopping Sequence Number* (HSN), and the TDMA *Frame Number* (FN); see Section 5.3.

The use of frequency hopping is an option left to the network operator, which can be decided on an individual cell basis. Therefore a mobile station must be able to switch to frequency hopping if a base station notices adverse conditions and decides to activate frequency hopping.

5.2.4 Summary

A physical GSM channel is defined by a sequence of frequencies and a sequence of TDMA frames. The RFCH sequence is defined by the frequency hopping parameters, and the temporal sequence of time slots of a physical channel is defined as a sequence of frame numbers and the time slot number within the frame. Frequencies for the uplink and downlink are always assigned as a pair of frequencies with a 45 MHz duplex separation.

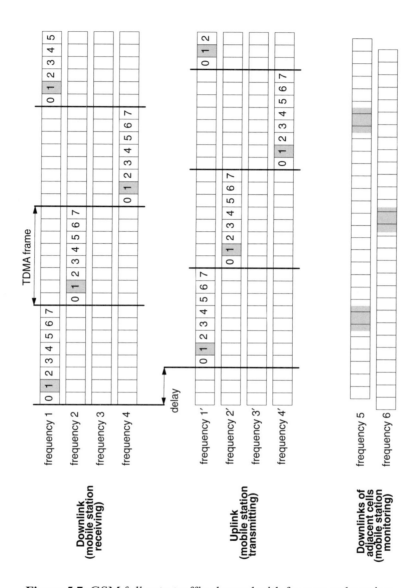

Figure 5.7: GSM full-rate traffic channel with frequency hopping

As shown above, GSM uses a series of parameters to define a specific physical channel of a base station. Summarizing, these parameters are

- *Mobile Allocation Index Offset* (MAIO)
- *Hopping Sequence Number* (HSN)
- *Training Sequence Code* (TSC)
- *Time Slot Number* (TN)
- *Mobile Allocation* (MA), also known as *RFCH Allocation*

- Type of logical channel carried on this physical channel
- The number of the logical subchannel (if used) – *Subchannel Number* (SCN)

Within a logical channel, there can be several subchannels (e.g. subrate multiplexing of the same channel type). The TDMA frame sequence can be derived from the type of the channel and the logical subchannel if present.

5.3   Synchronization

For the successful operation of a mobile radio system, synchronization between mobile stations and the base station is necessary. Two kinds of synchronization are distinguished: frequency synchrony and time synchrony of the bits and frames. Frequency synchronization is necessary so that transmitter and receiver frequencies agree. The objective is to compensate for tolerances of the less expensive and therefore less stable oscillators in the mobile stations by obtaining an exact reference from the base station and to follow it. Bit and frame synchrony are important in two regards for TDMA systems. First, the propagation time differences of signals from different mobile stations have to be adjusted, so that the transmitted bursts are received synchronously with the time slots of the base station and that bursts in adjacent time slots don't overlap and interfere with each other. Second, synchrony is needed for the frame structure since there is a higher-level frame structure superimposed on the TDMA frames for multiplexing logical signaling channels onto one physical channel. The synchronization procedures defined for GSM will be explained in the following section.

5.3.1   Frequency and Clock Synchronization

A GSM base station transmits signals on the frequency carrier of the *Broadcast Control Channel* (BCCH) which allow a mobile station to synchronize with the base station. Synchronization means on one hand the time-wise synchronization of mobile station and base with regard to bits and frames, and on the other hand tuning the mobile station to the correct transmitter and receiver frequencies.

For this purpose, the BTS provides the following signals (Figure 5.6):

- *Synchronization Channel* (SCH) with extra long *Synchronization Bursts* (SB), which facilitate synchronization
- *Frequency Correction Channel* (FCCH) with *Frequency Correction Bursts* (FB)

Because of the 0.3-GMSK modulation procedure used in GSM, a data sequence of logical "0" generates a pure sine wave signal, i.e. broadcasting of the FB corresponds to an unmodulated carrier (frequency channel) with a frequency shift of $1625/24$ kHz (≈ 67.7 kHz) above the nominal carrier frequency (Figure 5.8). In this way, the mobile station can keep exactly synchronized by periodically monitoring the FCCH. On the other hand, if the frequency of the BCCH is still unknown, it can search for the channel with the highest signal level. This channel is with all likelihood a BCCH channel, because dummy bursts must be transmitted on all unused time slots in this channel, whereas not all time slots are always used on other carrier frequencies. Using the FCCH sine wave signal allows identification of a BCCH and synchronization of a mobile station's oscillator.

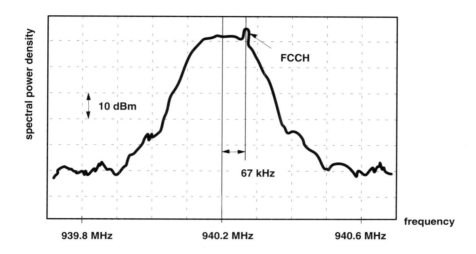

Figure 5.8: Typical power spectrum of a BCCH carrier

For the time synchronization, TDMA frames in GSM are cyclically numbered *modulo* 2 715 648 (=26 × 51 × 2^{11}) with the TDMA *Frame Number* (FN). One cycle generates the so-called hyperframe structure which comprises 2 715 648 TDMA frames. This long numbering cycle of TDMA frames is used to synchronize the ciphering algorithm at the air interface (see Section 6.3). Each base station BTS periodically transmits the *Reduced TDMA Frame Number* (RFN) on the *Synchronization Channel* (SCH). With each *Synchronization Burst* (SB) the mobiles thus receive information about the number of the current TDMA frame. This enables each mobile station to be time-synchronized with the base station.

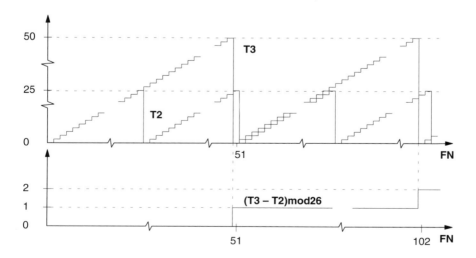

Figure 5.9: Values T2 and T3 for the calculation of RFN

The reduced TDMA frame number (RFN) has a length of 19 bits. It consists of three fields: T1 (11 bits), T2 (5 bits) and T3' (3 bits). These three fields are defined by (with div designating integer division):

$$T1 = FN \text{ div } (26 \times 51) \qquad [0 - 2047]$$
$$T2 = FN \text{ mod } 26 \qquad [0 - 25]$$
$$T3' = (T3 - 1) \text{ div } 10 \qquad [0 - 4]$$
$$\text{with } T3 = FN \text{ mod } 51 \qquad [0 - 50]$$

The sequences of running values of T2 and T3 are illustrated in Figure 5.9. The value crucial for the reconstruction of the frame number FN is the difference $(T3 - T2)$ between the two fields.

The time synchronization of a mobile station and its time slots, TDMA frames, and control channels is based on a set of counters which run continuously, independent of mobile or base station transmission. Once these counters have been started and correctly initialized, the mobile station is in a synchronized state with the base station.

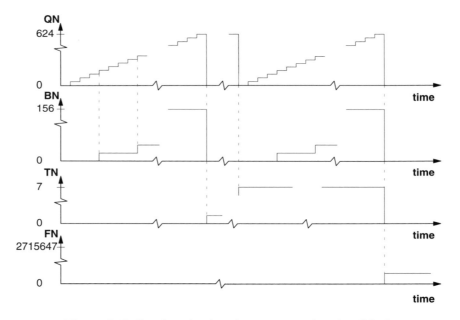

Figure 5.10: Synchronization timers, somewhat simplified:
the TDMA frame duration is 156.25 bit times

The following four counters are kept for this purpose:
- Quarter Bit Counter counting the *Quarter Bit Number* (QN)
- Bit Counter counting the *Bit Number* (BN)
- Time Slot Counter counting the *Time Slot Number* (TN)
- Frame Counter counting the *TDMA Frame Number* (FN)

Because of the bit and frame counting, these counters are of course interrelated, namely in such a way that the subsequent counter counts the overflows of the preceding counter. The following principle is used (Figure 5.10): QN is incremented every $12/13\,\mu s$; BN is obtained from it by integer division (BN = QN div 4). With each transition from 624 to 0 the time slot number TN is incremented, and each overflow of TN increments the frame counter FN by 1.

The timers can be reset and restarted when receiving a *Synchronization Burst* (SB). The Quarter Bit Counter is set by using the timing of the training sequence of the burst, whereas the *Time Slot Number* (TN) is reset to 0 with the end of the burst. The *TDMA Frame Number* (FN) can then be calculated from the *Reduced TDMA Frame Number* (RFN) transmitted on the *Synchronization Channel* (SCH):

$$FN = 51 \times ((T3 - T2) \bmod 26) + T3 + 51 \times 26 \times T1$$

$$\text{with } T3 = 10 \times T3' + 1$$

It is important to recalculate T3 from T3′, although, because of the binary representation, only the integer part of the division by 10 is taken into account.

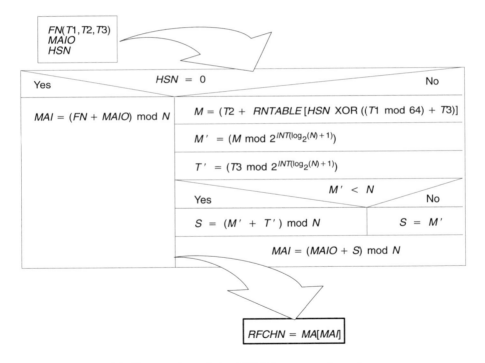

Figure 5.11: Generation of the GSM frequency hopping sequence

If the optional frequency hopping procedure is used (see Section 5.2.3), an additional mapping of the TDMA frame number onto the frequency to be used is required besides the evaluation of the synchronization signals from the FCCH and SCH. One has to obtain the index number of the frequency channel on which the current burst has to be transmitted from the *Mobile Allocation* (MA) table. This process uses a predefined RFNTABLE, the *Frame Number* (FN), and a *Hopping Sequence Number* (HSN); see Figure 5.11. The MA holds N frequencies, with a maximum value of 64 for N. With this procedure, every burst is sent on a different frequency in a cyclic way.

5.3.2 Adaptive Frame Synchronization

The mobile station can be anywhere within a cell, which means the distance between mobile and base station may vary. Thus the signal propagation times between mobile and base station vary. Due to the mobility of the subscribers, the bursts received at the base would be offset. The TDMA procedure cannot tolerate such time shifts, since it is based on the exact synchronization of transmitted and received data bursts. Bursts transmitted by different mobile stations in adjacent time slots must not overlap when received at the base by more than the guard period (Figure 5.6), even if the propagation times within the cell are very different. To avoid such collisions, the start of transmission time from the mobile station is advanced in proportion to the distance from the base station. The process of adapting the transmissions from the mobile stations to the TDMA frame is called *adaptive frame alignment*.

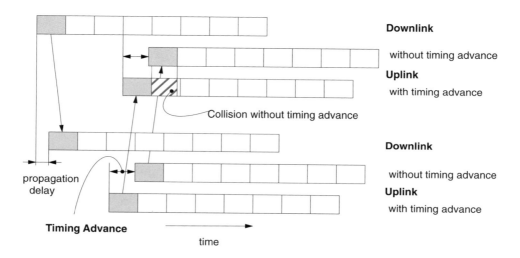

Figure 5.12: Operation of timing advance

For this purpose, the parameter *Timing Advance* (TA) in each SACCH Layer 1 protocol block is used (Figure 5.18). The mobile station receives from the base station on the SACCH downlink the TA value it must use; it reports the actually used value on the SACCH uplink. There are 64 steps for the timing advance which are coded as 0 to 63. One step corresponds to one bit period. Step 0 means no timing advance, i.e. the frames are transmitted with a time shift of 3 slots or 468.75 bit durations with regard to the downlink. At step 63, the timing of the uplink is shifted by 63 bit durations, such that the TDMA frames are transmitted on the uplink only with a delay of 405.75 bit durations. So the required adjustment always corresponds to twice the propagation time or is equal to the round-trip delay (Figure 5.12). In this way, the available range of values allows a compensation over a maximum propagation time of 31.5 bit periods ($\approx 113.3\,\mu s$). This corresponds to a maximum distance between mobile and base station of 35 km. A GSM cell may therefore have a maximum diameter of 70 km. The distance from the base station or the currently valid TA value for a mobile station is therefore an important handover criterion in GSM networks (see Section 8.4.3).

The adaptive frame alignment technique is based on continuous measurement of propagation delays by the base station and corresponding timing advance activity by the mobile station. In the case of an (unreserved) random access to the RACH, a channel must first be established. The base station has in this case not yet had the opportunity to measure the distance of the mobile station and to transmit a corresponding timing advance command. If a mobile station transmits an access burst in the current time slot, it uses a timing advance value of 0 or a default value. To minimize collisions with subsequent time slots at the base station, the access burst AB has to be correspondingly shorter than the time slot duration (Figure 5.13). This explains the long duration AB of the guard period of 68.25 bit periods, which can compensate for the propagation delay if a mobile station sends an access burst from the boundary of a cell of 70 km diameter.

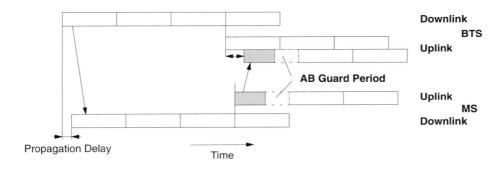

Figure 5.13: Timing for RACH random multiple access

5.4   Mapping of Logical Channels onto Physical Channels

The mapping of logical onto physical channels has two components: mapping in frequency and mapping in time. The mapping of a logical channel onto a physical channel in the frequency domain is based on the TDMA frame number FN, the frequencies allocated to base and mobile stations – *Cell Allocation* (CA) and *Mobile Allocation* (MA) – and the rules for the optional frequency hopping (see Section 5.2.3).

In the time domain, the logical channels are organized by the definition of complex superstructures on top of the TDMA frames, forming so-called multiframes, superframes and hyperframes (Figure 5.14). For the mapping of logical onto physical channels, we are interested in the multiframe domain. These multiframes allow us to map (logical) subchannels onto physical channels. Two kinds of multiframes are defined (Figure 5.15): a multiframe consisting of 26 TDMA frames (predominantly payload – speech and data – frames) and a multiframe of 51 TDMA frames (predominantly signaling frames).

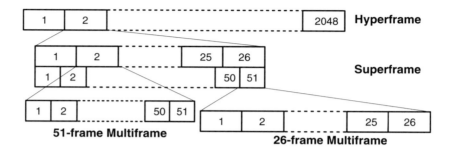

Figure 5.14: GSM frame structures

Each hyperframe is divided into 2048 superframes. With its long cycle period of 3 h 28 min 53.760 s, it is used for the synchronization of user data encryption. A superframe consists of 1326 consecutive TDMA frames which therefore lasts for 6.12 s, like 51 multiframes of 26 TDMA frames or 26 multiframes of 51 TDMA frames. These multiframes are again used to multiplex the different logical channels onto a physical channel as shown below.

Each 26 subsequent TDMA frames form a multiframe which multiplexes two logical channels, a *Traffic Channel* (TCH) and the *Slow Associated Control Channel* (SACCH), onto the physical channel (Figure 5.16). This process uses only one time slot per TDMA frame for the corresponding multiframe (e.g. time slot 3 in Figure 5.15), since a physical channel consists of just one time slot per TDMA frame. Besides the 24 TCH frames for user data, this multiframe also contains an AC frame for signaling data (SACCH data).

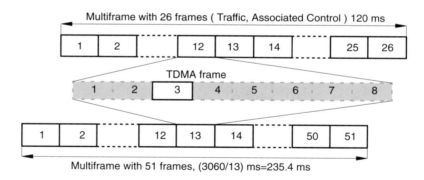

Figure 5.15: GSM multiframes

One frame (the 26th) remains unused in the case of a full-rate TCH (IDLE / AC); it is reserved for the introduction of two half-rate TCHs; then the 26th frame will be used to carry the SACCH channels of the other half. The data of the *Fast Associated Control Channel* (FACCH) is transmitted by occupying one half of the bits in eight consecutive bursts, by "stealing" these bits from the TCH. For this purpose, the *Stealing Flags* of the bursts are set (Figure 5.6).

A subscriber has available a gross data rate of 271 kbit/s ÷ 8 = 33.9 kbit/s (Section 5.2). Of this budget, 9.2 kbit/s are for signaling, synchronization, and guard periods of the burst. Of the remaining 24.7 kbit/s, in the case of the 26-frame multiframe, 22.8 kbit/s are left for the coded and enciphered user data of a full-rate channel, and 1.9 kbit/s remain for the SACCH and IDLE.

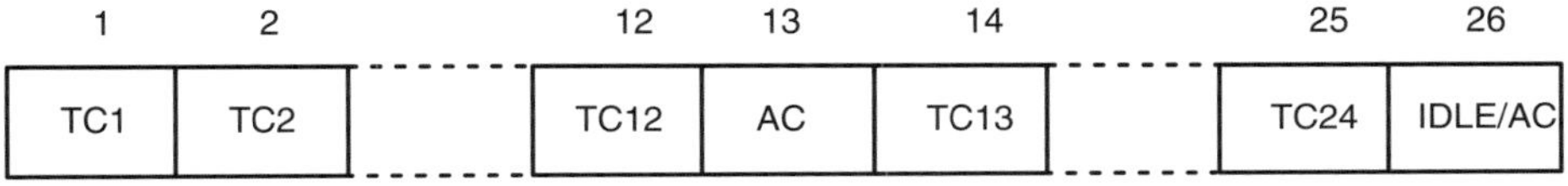

1	2		12	13	14		25	26
TC1	TC2		TC12	AC	TC13		TC24	IDLE/AC

Figure 5.16: Channel organization in a 26-frame multiframe

Each 51 consecutive time slots of a TDMA frame form a multiframe (Figure 5.6) to form the structure for the transmission of the control channels which are not associated with a TCH (all except FACCH and SACCH). According to channel configuration (Section 5.1), the multiframe is used differently. In each case, multiframes of 51 TDMA frames serve the purpose of mapping several logical channels onto a physical channel.

Furthermore, some of these control channels are unidirectional, which results in different structures for uplink and downlink. For some configurations, two adjacent multiframes are required to map all the logical channels. Some examples are illustrated in Figure 5.17. They correspond to the combinations B2, B3, and B4 in Table 5.3 whereas for channels SDCCH and SACCH some 4 or 8 logical subchannels have been defined (D0, D1, ..., A0, A1, ...). One of the frequency channels of the CA of a base station is used to broadcast synchronization data (FCCH and SCH) and the BCCH. Since the base station has to transmit in each time slot of the BCCH carrier to enable a continuous measurement of the BCCH carrier by the mobile station, a *Dummy Burst* (DB) is transmitted in all time slots with no traffic.

On time slot 0 of the BCCH carrier, only two combinations of logical channels may be transmitted, the combinations B2 or B3 from Table 5.3: (BCCH+CCCH+FCCH+SCH+SDCCH+SACCH or BCCH+CCCH+FCCH+SCH). No other time slot of the CA must carry this combination of logical channels.

As one can see in Figure 5.17, in the time slot 0 of the BCCH carrier of a base station (downlink) the frames $1, 11, 21, ...$ are FCCH frames, and the subsequent frames $2, 12, 22, ...$ form SCH frames. Frames 3, 4, 5, 6 of the 51-frame BCCH multiframe transport the appropriate BCCH information, whereas the remaining frames may contain different combinations of logical channels. Once the mobile station has synchronized by using the information from FCCH and SCH, it can determine from the information in the FCCH and SCCH how the remainder of the BCCH is constructed. For this purpose, the base station *Radio Resource Management* periodically transmits a set of messages to all mobile stations in this cell.

These *System Information Messages* comprise six types, of which only Types 1 through 4 are of interest here. Using the TDMA frame number (FN), one can determine which type is to be sent in the current time slot by calculating a *Type Code* (TC):

$$TC = (FN \text{ div } 51) \bmod 8$$

Table 5.5 shows how the type code TC determines the type of the system information message to be sent within the current multiframe.

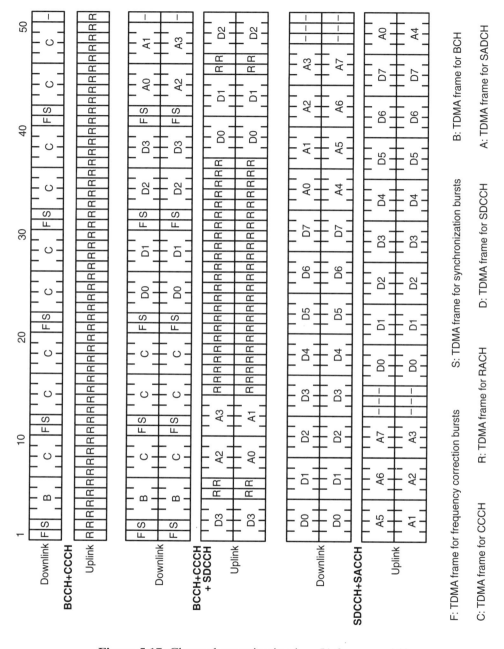

Figure 5.17: Channel organization in a 51-frame multiframe

Of the parameters contained in such a message, the following are of special interest: BS_CC_CHANS determines the number of physical channels which support a *Common Control Channel* (CCCH). The first CCCH is transmitted in time slot 0, the second one in time slot 2, the third one in time slot 4, and the fourth one in time slot 6 of the BCCH carrier. Another parameter, BS_CCCH_SDCCH_COMB, determines whether the *Dedi-*

cated Control Channels (DCCHs) SDCCH(0–3) and SACCH(0–3) are transmitted together with the CCCH on the same physical channel. In this case, each of these dedicated control channels consists of four subchannels.

Table 5.5: Mapping of frame number onto BCCH message

TC	System information message
0	Type 1
1	Type 2
2, 6	Type 3
3, 7	Type 4
4, 5	any (optional)

Each of the CCCHs of a base station is assigned a group CCCH_GROUP of mobile stations. Mobile stations are allowed random access (RACH) or receive paging information (PCH) only on the CCCH assigned to this group. Furthermore, a mobile station needs only to listen for paging information on every Nth block of the *Paging Channel* (PCH). The number N is determined by multiplying the number of paging blocks per 51-frame multiframe of a CCCH with the parameter BS_PA_MFRMS designating the number of multiframes between paging frames of the same *Paging Group* (PAGING_GROUP).

Especially in cells with high traffic, the CCCH and paging groups serve to subdivide traffic and to reduce the load on the individual CCCHs. For this purpose, there is a simple algorithm which allows each mobile station to calculate its respective CCCH_GROUP and PAGING_GROUP from its IMSI and parameters BS_CC_CHANS, BS_PA_MFRMS and N.

5.5   Radio Subsystem Link Control

The radio interface is characterized by another set of functions of which only the most important ones will be discussed in the following. One of these functions is the control of the radio link: *Radio Subsystem Link Control*, with the main activities of received-signal quality measurement (quality monitoring) for cell selection and handover preparation, and of transmitter power control.

If there is no active connection, i.e. if the mobile station is at rest, the *Base Station Subsystem* (BSS) has no tasks to perform. The MS, however, is still committed to continuously observing the BCCH carrier of the current and neighboring cells, so that it would be able to select the cell in which it can communicate with the highest probability. If a new cell needs to be selected, a *Location Update* may become necessary.

During a connection (TCH or SDCCH), the functions of channel measurement and power control serve to maintain and optimize the radio channel; this also includes adaptive frame alignment (Section 5.3.1) and frequency hopping (Section 5.2.3). Both need to be done until the current base can hand over the current connection to the next base station.

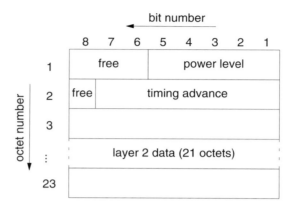

Figure 5.18: SACCH block format

These link control functions are performed over the SACCH channel. Two fields are defined in an SACCH block (Figure 5.18) for this purpose, the power level and the *Timing Advance* (TA). On the downlink, these fields contain values as assigned by the BSS. On the uplink, the MS inserts its currently used values. The quality monitoring measurement values are transmitted in the data part of the SACCH block.

The following illustrates the basic operation of the *Radio Subsystem Link Control* at the BSS side for an existing connection; the detailed explanation of the respective functions is given later. In principle, the radio link control can be subdivided into three tasks: measurement collection and processing, transmitter power control, and handover control.

In the example of Figure 5.19, the process BSS_Link_Control starts at initialization the processes BSS_Power_Control and BSS_HO_Control and then enters a measurement loop, which is only left when the connection is terminated. In this loop, measurement data is periodically received (every 480 ms) and current mean values are calculated. At first, these measurement data are supplied to the transmitter power control to adapt the power of MS and BS to a new situation if necessary. Thereafter, the measurement data and the result of the power control activity are supplied to the handover process, which can then decide whether a handover is necessary or not.

5.5.1 Channel Measurement

The task of *Radio Subsystem Link Control* in the mobile station includes identification of the reachable base stations and measurement of their respective received signal level and channel quality (quality monitoring task). In idle mode, these measurements serve to select the current base station, whose paging channel PCH is then periodically examined and on whose RACH desired connections can be requested.

During a connection, i.e. on a TCH or SDCCH with respective SACCH / FACCH, this measurement data is transmitted on the SACCH to the base station as a *measurement report / measurement info*. These reports serve as inputs for the handover and power control algorithms.

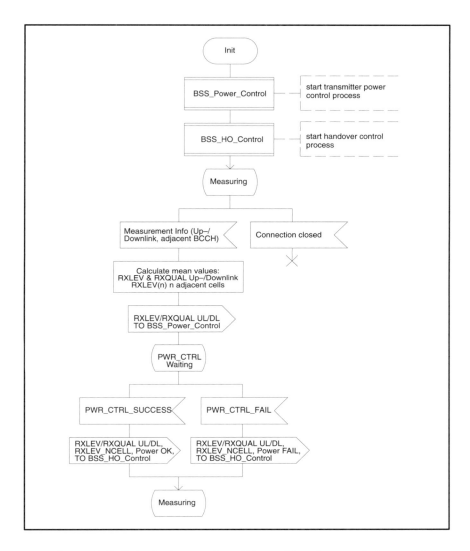

Figure 5.19: Principal operation of the *radio subsystem link control*

The measurement objects are on one hand the uplink and downlink of the current channel (TCH or SDCCH), and on the other hand the BCCH carriers which are continuously broadcast with constant power by all BTSs in all time slots. It is especially important to keep the transmitter power of the BCCH carriers constant to allow comparisons between neighboring base stations. A list of neighboring base stations' BCCH carrier frequencies, called the *BCCH Allocation* (BA) is supplied to each mobile by its current BTS, to enable measurement of all cells which are candidates for a handover. The cell identity is broadcast as the BSIC on the BCCH. Furthermore, up to 36 BCCH carrier frequencies and their BSICs can be stored on the SIM card.

In principle, GSM uses two parameters to describe the quality of a channel: the *Received Signal Level* (RXLEV), measured in dBm, and the *Received Signal Quality* (RXQUAL), measured as bit error ratio in per cent before error correction (Table 5.6 and Table 5.7).

The received signal power is measured continuously by mobile and base stations in each received burst within a range of −110 dBm to −48 dBm. The respective RXLEV values are obtained by averaging.

Table 5.6: Measurement range of the received signal level

Level	Received signal level	
	From	To
RXLEV_0	–	−110 dBm
RXLEV_1	−110 dBm	−109 dBm
.	.	.
.	.	.
.	.	.
RXLEV_62	−49 dBm	−48 dBm
RXLEV_63	−48 dBm	–

The bit error ratio before error correction can be determined in a variety of ways. For example, it can be estimated from information obtained from channel estimation for equalization from the training sequences, or the number of erroneous (corrected) bits can be determined through repeated coding of the decoded, error-corrected data blocks and comparison with the received data. Since the data before error correction is presented as blocks of 456 bits (see Section 6.2 and Figure 6.9), the bit error ratio can only be given with a quantizing resolution of 2×10^{-3}. Again, the value of RXQUAL is determined from this information by averaging.

Table 5.7: Measurement range of bit error ratio

Level	Bit error ratio	
	From	To
RXQUAL_0	–	0.2%
RXQUAL_1	0.2%	0.4%
RXQUAL_2	0.4%	0.8%
RXQUAL_3	0.8%	1.6%
RXQUAL_4	1.6%	3.2%
RXQUAL_5	3.2%	6.4%
RXQUAL_6	6.4%	12.8%
RXQUAL_7	12.8%	–

5.5.1.1 Channel Measurement during Idle Mode

In idle mode (see also Figure 7.17) the mobile station must always stay aware of its environment. The main purpose is to be able to assign a mobile station to a cell, whose BCCH carrier it can decode reliably. If this is the case, the mobile station is able to read system and paging information. If there is a desire to set up a connection, the mobile station can most likely communicate with the network.

There are two possible starting situations:

- The MS has no a priori knowledge about the network at hand, especially which BCCH carrier frequencies are in use.
- The MS has a stored list of BCCH carriers.

In the first case, the more unfavorable of the two, the mobile has to search through all the 124 GSM frequencies, measure their signal power level, and calculate an average from at least five measurements. The measurements of the individual carriers should be evenly distributed over an interval of three to five seconds. After at most five seconds, a minimum of 629 measurement values are available that allow the 124 RXLEV values to be determined. The carriers with the highest RXLEV values are very likely BCCH carriers, since continuous transmission is required on them. Final identification occurs with the frequency correction burst of the FCCH. Once the received BCCH carriers have been found, the mobile station starts to synchronize with each of them and reads the system information, beginning with the BCCH with the highest RXLEV value.

This orientation concerning the current location can be accelerated considerably, if a list of BCCH carriers has been stored on the SIM card. Then the mobile station tries first to synchronize with some known carrier. Only if it cannot find any of the stored BCCH carrier frequencies, does it start with the normal BCCH search. A mobile station can store several lists for the recently visited networks.

5.5.1.2   Channel Measurement during a Connection

During a traffic (TCH) or signaling (SDCCH) connection, the channel measurement of the mobile station occurs over an SACCH interval, which comprises 104 TDMA frames in the case of a TCH channel (480 ms) or 102 TDMA frames (470.8 ms) in the case of an SDCCH channel.

For the channel at hand, two parameters are determined: the received signal level RXLEV and the signal quality RXQUAL. These two values are averaged over a SACCH interval (480 or 470.8 ms) and transmitted to the base station on the SACCH as a *measurement report / measurement info.* This way the downlink quality of the channel assigned to the mobile station can be judged. In addition to these measurements of the downlink by the mobile station, the base station also measures the RXLEV and RXQUAL values of the respective uplink.

In order to make a handover decision, information about possible handover targets must be available. For this purpose, the mobile station has to observe continuously the BCCH carriers of up to six neighboring base stations. The RXLEV measurements of the neighboring BCCH carriers are performed during the mobile station's unused time slots (see Figure 5.7). The BCCH measurement results of the six strongest signals are included in the measurement report transmitted to the BSS.

However, the received signal power level and the frequency of a BCCH carrier alone are not a sufficient criterion for a successful handover. Because of the frequency reuse in cellular networks, and especially in the case of small clusters, it is possible that a cell can receive the same BCCH carrier from more than one neighboring cell, i.e. there exist several neighboring cells which use the same BCCH carrier. It is therefore necessary, to also know the identity (BSIC) of each neighboring cell. Simultaneously with the signal level measurement, the mobile station has to synchronize with each of the six neighboring BCCHs and read at least the SCH information.

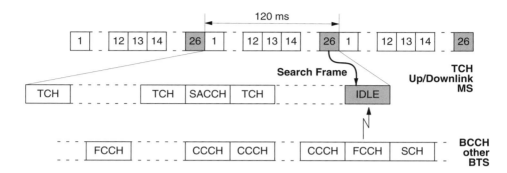

Figure 5.20: Synchronization with adjacent cells during a call

For this purpose, one must first search for the FCCH burst of the BCCH carrier; then the SCH can be found in the next TDMA frame. Since the FCCH / SCH / BCCH is always transmitted in time slot 0 of the BCCH carrier, the search during a conversation for FCCHs can only be conducted in unused frames, i.e. in case of a full-rate TCH in the IDLE frame of the multiframe (frame number 26 in Figure 5.16 and Figure 5.20). These free frames are therefore also known as *search frames*.

Therefore there are exactly four search frames within an SACCH block of 480 ms (four 26-frame multiframes of 120 ms). The mobile station has to examine the surrounding BCCH carriers for FCCH bursts, in order to synchronize with them and to decode the SCH. But how can one search for synchronization points exactly within these frames during synchronized operation?

This is possible because the actual traffic channel and the respective BCCH carriers use different multiframe formats. Whereas the traffic channel uses the 26-frame multiframe format, time slot 0 of the BCCH carrier with the FCCH / SCH / BCCH is carried on a 51-frame multiframe format. This ratio of the different multiframe formats has the effect that the relative position of the search frames (frame 26 in a TCH multiframe) is shifting with regard to the BCCH multiframe by exactly one frame each 240 ms (Figure 5.21). Figuratively speaking, the search frame is traveling along the BCCH multiframe in such a way that at most after 11 TCH multiframes (=1320 ms) a frequency correction burst of a neighboring cell becomes visible in a search frame.

In this way, the mobile station is able to determine the BSIC for the respective RXLEV measurement value. Only BCCH carrier measurements whose identity can be established without doubt are included in the measurement report to the base station.

The base station can now make a handover decision based on these values, on the distance of the mobile station, and on the momentary interference of unused time slots.

The algorithm for handover decisions has not been included in the GSM standard. The network operators may use algorithms which are optimized for their network or the local situation. GSM only gives a basic proposal which satisfies the minimum requirements for a handover decision algorithm. This algorithm defines threshold values, which must be violated in one or the other direction to arrive at a safe handover decision and to avoid so-called ping-pong handovers, which oscillate between two cells. Although the decision algorithm is part of *Radio Subsystem Link Control*, its discussion is postponed and will be treated together with handover signaling (see Section 8.4.3).

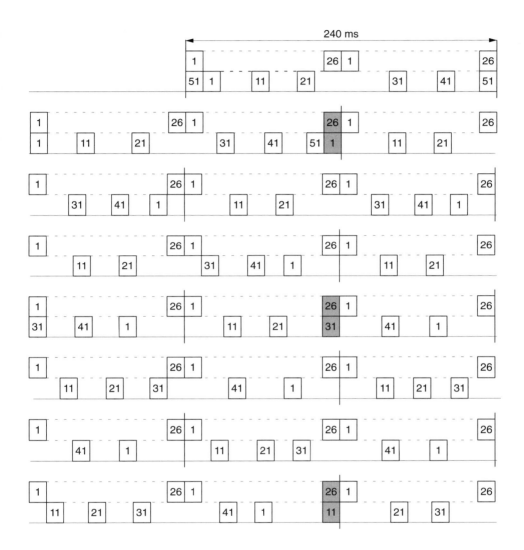

Figure 5.21: Principle of FCCH search during the search frame

5.5.2 Transmission Power Control

Power classes (Table 5.8) are used for classification of base and mobile stations. The transmission power can also be controlled adaptively. As part of the *Radio Subsystem Link Control*, the mobile stations transmitter power is controlled in steps of 2 dBm.

The GSM transmitter power control has the purpose of limiting the mobile station's transmitter power to the minimum necessary level, in such a way that the base station receives signals from different mobile stations at approximately the same power level. Sixteen power control steps are defined for this purpose: Step 0 (43 dBm = 20 W) to Step 15 (13 dBm). Starting with the lowest, Step 15, the base station can increment the transmitter power of the mobile station in steps of 2 dBm up to the maximum power level of the respective power class of the mobile station. Similarly, the transmitter power of the base station can be controlled in steps of 2 dBm, with the exception of the BCCH carrier of the base station, which must remain constant to allow comparative measurements of neighboring BCCH carriers by the mobile stations.

Table 5.8: GSM power classes

Power class	Max. peak transmission power	
	Mobile station	Base station
1	20 W (43 dBm)	320 W
2	8 W (39 dBm)	160 W
3	5 W (37 dBm)	80 W
4	2 W (33 dBm)	40 W
5	0.8 W (29 dBm)	20 W
6	−	10 W
7	−	5 W
8	−	2.5 W

Transmission power control is based on the measurement values RXLEV and RXQUAL, for which one has defined upper and lower thresholds for uplink and downlink (Table 5.9). Network management defines the adjustable parameters P and N. If the values of P for the last N calculated mean values of the respective criterion (RXLEV or RXQUAL) are above or below the respective threshold value, the BSS can adjust the transmitter power (Figure 5.22).

Table 5.9: Thresholds for transmitter power control

Threshold parameter	Typical value	Meaning
L_RXLEV_UL_P	−103 to −73 dBm	
L_RXLEV_DL_P	−103 to −73 dBm	Threshold for raising of transmission power
L_RXQUAL_UL_P	−	in uplink or downlink
L_RXQUAL_DL_P	−	
U_RXLEV_UL_P	−	
U_RXLEV_DL_P	−	Threshold for reducing of transmission
U_RXQUAL_UL_P	−	power in uplink or downlink
U_RXQUAL_DL_P	−	

If the thresholds U_xx_UL_P of the uplink are exceeded, the transmission power of the mobile station is reduced; in the other case, if the signal level is below the threshold L_xx_UL_P, the mobile station is ordered to increase its transmitter power. In an analogous way, the transmitter power of the base station can be adjusted, when the criteria for the downlink are exceeded in either direction.

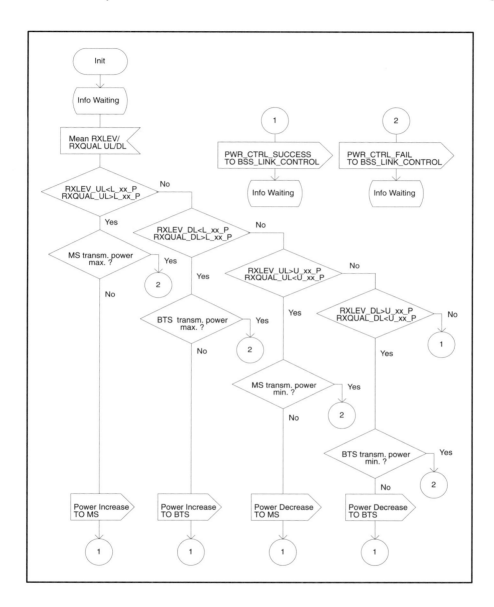

Figure 5.22: Schematic operation of transmitter power control

Even if the mobile or base station signal levels stay within the thresholds, the current RXLEV / RXQUAL values can cause a change to another channel of the same or another cell based on the handover thresholds (Table 8.1).

For this reason, checking for transmitter thresholds is immediately followed by a check of the handover thresholds as the second part of the *Radio Subsystem Link Control* (Figure 5.19 and Figure 8.17).

If one of the threshold values is exceeded in either direction and the transmitter power cannot be adjusted accordingly, i.e. the respective transmitter power has reached its maximum or minimum value, this is an overriding cause for handover (PWR_CTRL_FAIL, see Table 8.2) which the BSS must communicate immediately to the MSC (see Section 8.4).

5.5.3 Disconnection due to Radio Channel Failure

The quality of a radio channel can vary considerably during an existing connection, or it can even fail in the case of shadowing. This should not lead to immediate disconnection, since such failures are often of short duration. For this reason GSM has a special algorithm within the *Radio Subsystem Link Control* which continuously checks for connectivity. It consists of recognizing a *radio link failure* by the inability to decode signaling information on the SACCH. This connectivity check is done both in the mobile as well as in the base station. The connection is not immediately terminated, but is delayed so that only repeated consecutive failures (erroneous messages) represent a valid disconnect criterion.

On the downlink, the mobile station must check the frequency of erroneous, nondecodable messages on the SACCH. The error protection on the SACCH has very powerful error correction capabilities and thus guarantees a very low probability of 10^{-10} for nonrecognized, wrongly corrected bits in SACCH messages.

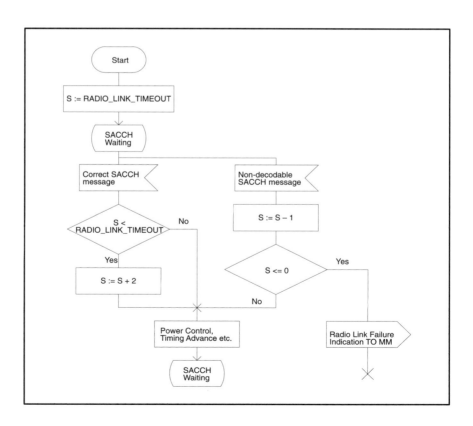

Figure 5.23: MS disconnect procedure

In this way, erroneous SACCH messages supply a measure for the quality of the downlink, which is already quite low when errors on the SACCH cannot be corrected any more. If a consecutive number of SACCH messages is erroneous, the link is considered bad, and the connection is terminated. For this purpose, a counter S has been defined which is incremented by 2 with each arrival of an error-free message, and decremented by 1 for each erroneous SACCH message (Figure 5.23). When the counter reaches the value $S=0$, the downlink is considered as failing, and the connection is terminated. This failure is signaled to the upper layers, *Mobility Management* (MM), which can start a *call reestablishment* procedure. The maximum value RADIO_LINK_TIMEOUT for the counter S therefore determines the interval length during which a channel has to fail before a connection is terminated. After assignment of a dedicated channel (TCH or SDCCH), the mobile station starts the checking process by initializing the counter S with this value (Figure 5.23), which can be set individually per cell and is broadcast on the BCCH.

The corresponding checks are also conducted on the uplink. In both cases, however, this requires continuous transmission of data on the SACCH, i.e. when no signaling data has to be sent, filling data is transmitted. On the uplink, current measurement reports are transmitted, whereas the downlink carries system information of Type 5 and Type 6 (see also Section 7.4.3).

5.5.4   Cell Selection and Operation in Power Conservation Mode

5.5.4.1   Cell Selection and Cell Reselection

A mobile station in idle mode must periodically measure the receivable BCCH carriers of the base stations in the area and calculate mean values RXLEV(n) from this data (see Section 5.5.1.1). Based on these measurements, the mobile station selects a cell, namely the one with the best reception, i.e. the mobile station is committed to this cell. This is called "camping" on this cell. In this state, accessing a service becomes possible, and the mobile station listens periodically to the paging channel (PCH).

Two criteria are defined for the automatic selection of cells: the path loss criterion C1 and the reselection criterion C2. The path loss criterion serves to identify cell candidates for camping. For such cells, C1 has to be greater than zero. At least every five seconds, a mobile station has to recalculate C1 and C2 for the current and neighboring cells. If the path loss criterion of the current cell falls below zero, the path loss to the current base station has become too large. A new cell has to be selected, which requires use of the criterion C2. If one of the neighboring cells has a value of C2 greater than zero, it becomes the new current cell.

The cell selection algorithm uses two further threshold values, which are broadcast on the BCCH:

- the minimum received power level RXLEV_ACCESS_MIN (typically -98 dBm to -106 dBm) required for registration into the network of the current cell
- the maximum allowed transmitter power MS_TXPWR_CCH (typically 31 dBm to 39 dBm) allowed for transmission on a control channel (RACH) before having received the first power control command

In consideration of the maximal transmitter power P of a mobile station, the *Path Loss Criterion* C1 is now defined using the minimal threshold RXLEV_ACCESS_MIN for network access and the maximal allowed transmitter power MS_TXPWR_MAX_CCH:

$$C1(n) = (\text{ RXLEV}(n) - \text{RXLEV\_ACCESS\_MIN}$$
$$- \text{maximum}(0, (\text{MS\_TXPWR\_MAX\_CCH} - P)))$$

The values of the path loss criterion C1 are determined for each cell for which a value RXLEV(n) of a BCCH carrier can be obtained. The cell with the lowest path loss can thus be determined using this criterion. It is the cell for which C1>0 has the largest value. During cell selection, the mobile station is not allowed to enter power conservation mode (DTX, see Section 5.5.4.2).

A prerequisite for cell selection is that the cell considered belongs to the home PLMN of the mobile station or that access to the PLMN of this cell is allowed. Beyond that, a *Limited Service Mode* has been defined with restricted service access, which still allows emergency calls if nothing else. In limited service mode, a mobile station can be camping on any cell but can only make emergency calls. Limited service mode exists when there is no SIM card in the mobile station, when the IMSI is unknown in the network or the IMEI is barred from service, but also if the cell with the best value of C1 does not belong to an allowed PLMN.

Once a mobile station is camping on a cell and is in idle mode, it should keep observing all the BCCH carriers whose frequencies, the *BCCH Allocation* (BA), are broadcast on the current BCCH. Having left idle mode, e.g. if a TCH has been assigned, the mobile station monitors only the six strongest neighboring BCCH carriers. A list of these six strongest neighboring BCCH carriers has already been prepared and kept up to date in idle mode. The BCCH of the camped-on cell must be decoded at least every 30 seconds. At least once every 5 minutes, the complete set of data from the six strongest neighboring BCCH carriers has to be decoded, and the BSIC of each of these carriers has to be checked every 30 seconds. This allows the mobile station to stay aware of changes in its environment and to react appropriately. In the worst case, conditions have changed so much that a new cell to camp on needs to be selected (cell reselection).

For this cell reselection, a further criterion C2, the *Reselection Criterion*, has been defined:

$$C2(n) = C1(n) + \text{CELL\_RESELECT\_OFFSET}$$
$$- (\text{TEMPORARY\_OFFSET} \times H(\text{PENALTY\_TIME} - T))$$

$$\text{with } H(x) = \begin{cases} 0 & \text{for } x < 0 \\ 1 & \text{for } x \geq 0 \end{cases}$$

The interval T in this criterion is the time passed since the mobile station observed the cell n for the first time with a value of C1>0. It is set back to 0 when the path loss criterion C1 falls to C1 < 0. The parameters CELL_RESELECT_OFFSET, TEMPORARY_OFFSET, and PENALTY_TIME are announced on the BCCH. But as a default, they are set to 0. Otherwise, the criterion C2 introduces a time hysteresis for cell reselection. It tries to ensure that the mobile station is camping on the cell with the highest probability of successful communication.

One exception for cell reselection is the case when a new cell belongs to another location area. In this case C2 must not only be larger than zero, but C2 > CELL_RESELCT_HYSTERESIS to avoid too frequent location updates.

5.5.4.2   Discontinuous Reception

To limit power consumption in idle mode and thus increase battery life in standby mode, the mobile station can activate the *Discontinuous Reception* (DRX) mode. In this mode, the receiver is turned on only for the phases of receiving paging messages and is otherwise in the power conservation mode which still maintains synchronization with BCCH signals through internal timers. In this DRX mode, measurement of BCCH carriers is performed only during unused time slots of the paging blocks.

5.6   Power-up Scenario

At this point, all the functions, protocols and mechanisms of the GSM radio interface have been presented which are needed to illustrate a basic power-up scenario. The following describes the basic events that occur during a power up of the mobile station. The scenario can be divided into several steps:

1. Provided a SIM card is present, immediately after turning on power, a mobile station starts the search for BCCH carriers. Normally, the station has a stored list of up to 32 carriers (Figure 5.24) of the current network. Signal level measurements are done on each of these frequencies (RXLEV). Alternatively, if no list is available, all GSM frequencies have to be measured to find potential BCCH carriers. Using the path loss criterion C1 and the threshold values stored with the list of carriers (RXLEV_ACCESS_MIN, MS_TXPWR_MAX_CCH), a first ordering can be done.

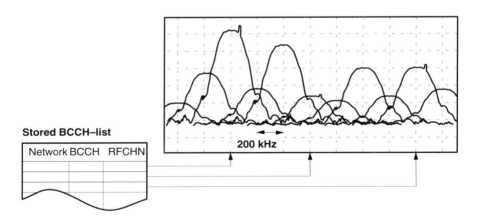

Figure 5.24: BCCH search in the power density spectrum (schematic)

2. After having found potential candidates based on the received signal level RXLREV, each carrier is investigated for the presence of an FCCH signal, beginning with the strongest signal. Its presence identifies the carrier as a BCCH carrier for synchronization. Using the sine wave signal allows coarse time synchronization as well as fine tuning of the oscillator.

3. The synchronization burst SB of the SCH in the TDMA frame immediately following the FCCH burst (Figure 5.17) has a long training sequence of 64 bits
   (Figure 5.6) which is used for fine tuning of the frequency correction and time synchronization. This way the mobile station is able to read and decode synchronization data from the SCH, the BSIC and the reduced TDMA frame number (RFN).
   This process starts with the strongest of all BCCH carriers. If a cell is identified
   using BSIC and path loss criterion C1, the cell is selected for camping on it.

4. The exact channel configuration of the selected cell is obtained from the BCCH
   data as well as the frequencies of the neighboring cells. Now the paging channel
   (PCH) of the current cell can be monitored, and the signal levels of the neighboring
   cells can be monitored. A list of the six strongest BCCH carriers can be established.

5. The mobile station must now prepare synchronization with the six cells with the
   strongest signal level (RXLEV) and read out their BCCH / SCH information, i.e.
   steps 1 to 4 above are to be performed continuously for the six neighboring cells with
   the best RXLEV values.

6. If significant changes are noticed using the path loss criterion C1 and the reselection criterion C2, the mobile station can start reselection of a new cell. Both criteria
   are determined periodically for the current BCCH and the six strongest neighbors.

To limit power consumption and to extend standby time of the battery, the mobile station
can activate the *Discontinuous Reception* (DRX) mode.

6 Coding, Authentication, and Ciphering

6.1 Source Coding and Speech Processing

Source coding reduces redundancy in the speech signal and thus results in signal compression, which means that a significantly lower bit rate is achieved than needed by the original speech signal. The speech coder / decoder is the central part of the GSM speech processing function, both at the transmitter (Figure 6.1) as well as at the receiver (Figure 6.2). The functions of the GSM speech coder and decoder are usually combined in one building block called the codec (COder/DECoder).

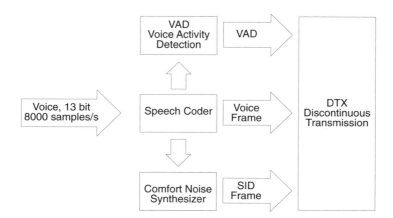

Figure 6.1: Schematic representation of speech functions at the transmitter

The analog speech signal at the transmitter is sampled at a rate of 8000 samples per second, and the samples are quantized with a resolution of 13 bits. This corresponds to a bit rate of 104 kbit/s for the speech signal. At the input to the speech codec, a speech frame containing 160 samples of 13 bits arrives every 20 ms. The speech codec compresses this speech signal into a source-coded speech signal of 260-bit blocks at a bit rate of 13 kbit/s. Thus the GSM speech coder achieves a compression ratio of 1 to 8. The source coding procedure will be briefly explained in the following; detailed discussions of speech coding procedures are given in [29].

A further ingredient of speech processing at the transmitter is the recognition of speech pauses, called *Voice Activity Detection* (VAD). The voice activity detector (Figure 6.2) decides, based on a set of parameters delivered by the speech coder, whether the current speech frame (20 ms) contains speech or a speech pause. This decision is used to turn off the transmitter amplifier during speech pauses, under control of the DTX block.

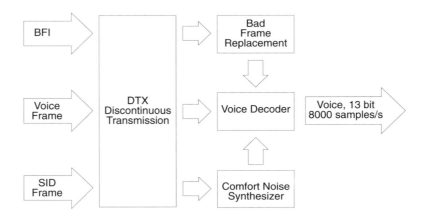

Figure 6.2: Schematic representation of speech functions at the receiver

The *Discontinuous Transmission* (DTX) mode takes advantage of the fact, that during a normal telephone conversation, both parties rarely speak at the same time, and thus each directional transmission path has to transport speech data only half the time. In DTX mode, the transmitter is only activated when the current frame indeed carries speech information. This decision is based on the VAD signal of speech pause recognition. On one hand, the DTX mode can reduce the power consumption and hence prolong the battery life; on the other hand, the reduction of transmitted energy also reduces the level of interference and thus improves the spectral efficiency of the GSM system. The missing speech frames are replaced at the receiver by a synthetic background noise signal called *Comfort Noise* (Figure 6.2). The parameters for the *Comfort Noise Synthesizer* are transmitted in a special SID frame.

This *Silence Descriptor* (SID) is generated at the transmitter from continuous measurements of the (acoustic) background noise level. It represents a speech frame which is transmitted at the end of a speech burst, i.e. at the beginning of a speech pause. In this way, the receiver recognizes the end of a speech burst and can activate the comfort noise synthesizer with the parameters received in the SID frame. The generation of this artificial background noise prevents that in DTX mode the audible background noise transmitted with normal speech bursts suddenly drops to a minimal level at a speech pause. This modulation of the background noise would have a very disturbing effect on the human listener and would significantly deteriorate the subjective speech quality. Insertion of comfort noise is a very effective countermeasure to compensate for this so-called noise-contrast effect.

Another loss of speech frames can occur, when bit errors caused by a noisy transmission channel cannot be corrected by the channel coding protection mechanism, and the block is received at the codec as a speech frame in error, which must be discarded. Such bad speech frames are flagged by the channel decoder with the *Bad Frame Indication* (BFI). In this case, the respective speech frame is discarded and the lost frame is replaced by a speech frame which is predictively calculated from the preceding frame. This technique is called *Error Concealment*. Simple insertion of comfort noise is not allowed. If 16 consecutive speech frames are lost, the receiver is muted to acoustically signal the temporary failure of the channel.

The speech compression proper takes place in the speech coder. The GSM speech coder uses a procedure known as *Regular Pulse Excitation − Long-Term Prediction − Linear Predictive Coder* (RPE-LTP). This procedure belongs to the family of hybrid speech coders. This hybrid procedure transmits part of the speech signal as the amplitude of a signal envelope, a pure wave form encoding, whereas the remaining part is encoded into a set of parameters. The receiver reconstructs these signal parts through speech synthesis (vocoder technique). Examples of envelope encoding are *Pulse Code Modulation* (PCM) or *Adaptive Delta Pulse Code Modulation* (ADPCM). A pure vocoder procedure is *Linear Predictive Coding* (LPC). The GSM procedure RPE-LTP as well as *Code Excited Linear Predictive Coding* (CELP) represent mixed (hybrid) approaches [11][29][32].

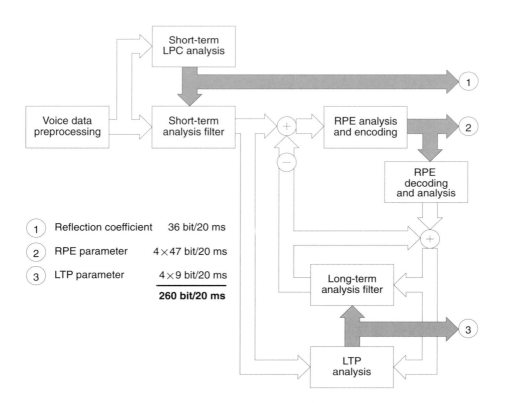

Figure 6.3: Simplified block diagram of the GSM speech coder

A simplified block diagram of the RPE-LTP coder is shown in Figure 6.3. Speech data generated with a sampling rate of 8000 samples per second and 13 bit resolution arrive in blocks of 160 samples at the input of the coder. The speech signal is then decomposed into three components: a set of parameters for the adjustment of the short-term analysis filter (LPC) also called *reflection coefficients*; an excitation signal for the RPE part with irrelevant portions removed and highly compressed; and finally a set of parameters for the control of the LTP long-term analysis filter. The LPC and LTP analyses supply 36 filter parameters for each sample block, and the RPE coding compresses the sample block to 188 bits of RPE parameters. This results in the generation of a frame of 260 bits every 20 ms, equivalent to a 13 kbit/s GSM speech signal rate.

The speech data preprocessing of the coder (Figure 6.3) removes the DC portion of the signal if present and uses a preemphasis filter to emphasize the higher frequencies of the speech spectrum. The preprocessed speech data is run through a nonrecursive lattice filter (LPC filter, Figure 6.3) to reduce the dynamic range of the signal. Since this filter has a "memory" of about 1 ms, it is also called short-term prediction filter. The coefficients of this filter, called reflection coefficients, are calculated during LPC analysis and transmitted in a logarithmic representation as part of the speech frame, *Log Area Ratios* (LARs).

Further processing of the speech data is preceded by a recalculation of the coefficients of the long-term prediction filter (LTP analysis in Figure 6.3). The new prediction is based on the previous and current blocks of speech data. The resulting estimated block is finally subtracted from the block to be processed, and the resulting difference signal is passed on to the RPE coder.

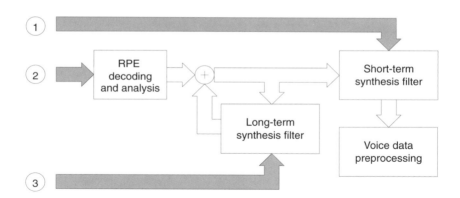

Figure 6.4: Simplified block diagram of the GSM speech decoder

After LPC and LTP filtering, the speech signal has been redundancy reduced, i.e. it already needs a lower bit rate than the sampled signal; however, the original signal can still be reconstructed from the calculated parameters. The irrelevance contained in the speech signal is reduced by the RPE coder. This irrelevance represents speech information that is not needed for the understandability of the speech signal, since it is hardly noticeable to human hearing and thus can be removed without loss of quality. On one hand, this results in a significant compression (factor $160 \times 13 / 188 \approx 11$); on the other hand, it has the effect that the original signal cannot be reconstructed uniquely. Figure 6.4 summarizes the reconstruction of the speech signal from RPE data, as well as the long-term and short-term synthesis from LTP and LPC filter parameters. In principle, at the receiver site, the functions performed are the inverse of the functions of the encoding process.

The irrelevance reduction only minimally affects the subjectively perceived speech quality, since the main objective of the GSM codec is not just the highest possible compression but also good subjective speech quality. To measure the speech quality in an objective manner, a series of tests were performed on a large number of candidate systems and competing codecs.

The base for comparison used is the *Mean Opinion Score* (MOS), ranging from MOS=1, meaning quality is very bad or unacceptable, to MOS=5, quality very good, fully acceptable. A series of coding procedures were discussed for the GSM system; they were examined in extensive hearing tests for their respective subjective speech quality [11]. Table 6.1 gives an overview of these test results; it includes as reference also ADPCM and frequency-modulated analog transmission. The GSM codec with the RPE-LTP procedure generates a speech quality with an MOS value of about 4 for a wide range of different inputs.

Table 6.1: MOS results of codec hearing tests [11]

CODEC	Process	Bit rate [kbit/s]	MOS
FM	Frequency Modulation	–	1.95
SBC-ADPCM	Subband-CODEC – Adaptive Delta-PCM	15	2.92
SBC-APCM	Subband-CODEC – Adaptive PCM	16	3.14
MPE-LTP	Multi-Pulse Excited LPC-CODEC – Long Term Prediction	16	3.27
RPE-LPC	Regular-Pulse Excited LPC-CODEC	13	3.54
RPE-LTP	Regular Pulse Excited LPC-CODEC – Long Term Prediction	13	≈ 4
ADPCM	Adaptive Delta Modulation	32	≥ 4

6.2   Channel Coding

6.2.1   Overview

The heavily varying properties of the mobile radio channel (see Section 2.1) result in an often very high bit error ratio, on the order of 10^{-3} to 10^{-1}. The highly compressed, redundancy-reduced source coding makes speech communication with acceptable quality nearly impossible; moreover, it makes reasonable data communication impossible. Suitable error correction procedures are therefore necessary to reduce the bit error probability into an acceptable range of about 10^{-5} to 10^{-6}. Channel coding, in contrast to source coding, adds redundancy to the data stream to enable detection and correction of transmission errors. It is only the modern high-performance coding and error correction techniques which essentially enable the implementation of a digital mobile communication system.

The GSM system uses a combination of several procedures: besides a block code, which generates parity bits for error detection, a convolutional code generates the redundancy needed for error correction. Furthermore, sophisticated interleaving of data over several blocks reduces the damage done by burst errors.

The individual steps of channel coding are shown in Figure 6.5:

- Calculation of parity bits (block code) and addition of fill bits
- Error protection coding through convolutional coding
- Interleaving

Finally, the coded and interleaved blocks are enciphered, distributed across bursts, modulated and transmitted on the respective carrier frequencies.

First the data is combined into blocks, partially supplemented by parity bits (depending on the logical channel), and then complemented to a block size suitable for the convolutional coder. This involves appending zero bits at the end of each data block, which allow a defined resetting procedure of the convolutional coder and thus a correct decoding decision. Finally, these blocks are run through the convolutional coder. The ratio of uncoded to coded block length is called the "rate" of the convolutional coder. Some of the redundancy bits generated by the convolutional coder are canceled again for some of the logical channels, i.e. the convolutional code is punctured. Puncturing increases the rate of the convolutional code,so it reduces the redundancy per block to be transmitted, and lowers the bandwidth requirements, such that the convolution-encoded signal fits into the available channel bit rate. The convolution-encoded bits are passed to the interleaver, which shuffles various bit streams. At the receiving site, the respective inverse functions are performed: deinterleaving, convolutional decoding, parity checking. Depending on the position within the transmission chain (Figure 6.5), one distinguishes between external error protection (block code) and internal protection (convolutional code).

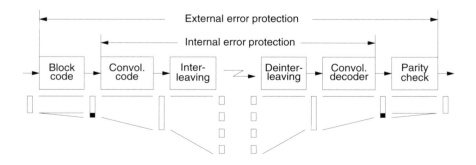

Figure 6.5: Stages of channel coding

The GSM channel coding will now be presented according to these stages. The error protection measures have different parameters depending on channel and type of transported data. Table 6.2 gives an overview. Note that the tail bits indicated in column 2 are the fill bits needed by the decoding process; they should not be confused with the tail bits of the bursts (see Section 5.2).

Table 6.2: Error protection coding and interleaving of logical channels

Channel type	Abbr.	Bit per block			Convol. code rate	Encoded bits per block	Interleav. depth
		Data	Parity	Tail			
TCH, full-rate	TCH/FS					456	8
Voice Class I		182	3	4	1/2	(378)	
Voice Class II		78	0	0	–	(78)	
TCH, full-rate, 9.6 kbit/s	TCH/F9.6	4×60	0	4	244/456	456	19
TCH, full-rate, 4.8 kbit/s	TCH/F4.8	60	0	16	1/3	228	19
TCH, half-rate, 4.8 kbit/s	TCH/H4.8	4×60	0	4	244/456	456	19
TCH, full-rate, 2.4 kbit/s	TCH/F2.4	72	0	4	1/6	456	8
TCH, half-rate, 2.4 kbit/s	TCH/H2.4	72	0	4	1/3	228	19
FACCHs	FACCH	184	40	4	1/2	456	8
SDCCHs, SACCHs	SDCCH, SACCH	184	40	4	1/2	456	4
BCCH, AGCH, PCH	BCCH, AGCH, PCH	184	40	4	1/2	456	4
RACH	RACH	8	6	4	1/2	36	1
SCH	SCH	25	10	4	1/2	78	1

The basic unit for all coding procedures is the data block. Depending on the logical channel, the length of the data block is different; however, after convolutional coding at the latest, data from all channels are transformed into units of 456 bits. Such a block of 456 bits transports a complete speech frame or a protocol message in most of the signaling channels, except for the RACH and SCH channels. The starting points are the blocks delivered to the input of the channel coder from the protocol processing in higher layers (Figure 6.6). A block from the speech codec consists of 260 bits of speech data. It is divided into two sections which have different sensitivity against bit errors, as a result of the different importance of the classes of speech bits generated by the speech codec. Class I includes bits which are more sensitive against interference (more impact on speech quality) and hence must be better protected. Speech bits of Class II, however, are less sensitive to transmission errors. They are therefore transmitted without convolutional coding, but are included in the interleaving process. The individual sections of a speech frame are therefore protected to differing degrees against transmission errors (unequal error protection). Blocks of traffic channels for data services have a length of N0 bits, the value of N0 being a function of the data service bit rate. The data streams of most of the signaling channels are constructed of blocks of 184 bits each; with the exception of the RACH and SCH which supply blocks of length P0 to the channel coder. The block length of 184 bits results from the fixed length of the protocol message frames of 23 octets on the signaling channels. The channel coding process maps pairs of subblocks of 57 bits onto the bursts such that it can fill a normal data burst NB (Figure 5.6).

6.2.2   External Error Protection: Block Coding

The block coding stage in GSM has the purpose of generating the parity bits for a block of data which allow the detection of errors in this block. In addition, these blocks are supplemented by fill bits (tail bits) to a block length suitable for further processing. Since block coding is the first or external stage of channel coding, the block code is also known as *external protection*. Figure 6.6 gives a brief overview showing which codes are used for which channels. In principle, only two kinds of codes are used: a *Cyclic Redundancy Check* (CRC) and a *fire code*.

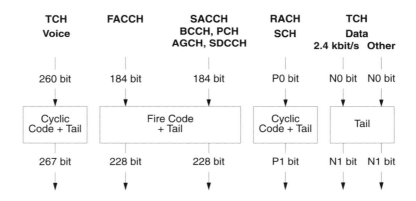

Figure 6.6: Overview of block coding for logical channels

6.2.2.1   Block Coding for Speech Traffic Channels

The most common form of a traffic channel is the speech channel. Speech data occurs on the TCH in speech frames (blocks) of 260 bits. Of these, 182 bits belong to Class I and are error-protected, whereas the remaining 78 bits belong to Class II and are not protected. A 3-bit *Cyclic Redundancy Check* (CRC) code is calculated for the first 50 bits of Class I. The generating polynomial for this CRC is

$$G_{CRC}(x) = x^3 + x + 1$$

Since cyclic codes are easily generated with a feedback shift register, they are often defined directly with this register representation (Figure 6.7). For initialization, the register is primed with the first three bits of the data block. The other data are shifted in bitwise into the feedback shift register; after the last data bit has been shifted out of the register, the register contains the check sum bits, which are then appended to the block.

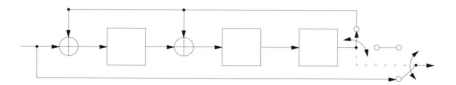

Figure 6.7: Feedback shift register for CRC

The operation of this shift register can be easily explained, if the bit sequences are also represented as polynomials like the generating function. The first 50 bits of a speech frame $D_0, D_1, ..., D_{49}$ are denoted as

$$D(x) = D_{49} x^{49} + D_{48} x^{48} + ... + D_1 x + D_0$$

If this data sequence is shifted through the shift register of Figure 6.7, after the register was primed with D_{47}, D_{48}, D_{49} followed by 50 shift operations, then the check sum bits $R(x)$ correspond to the remainder, which is left by dividing the data sequence $x^3 D(x)$ (supplemented by three zero bits) by the generating polynomial:

$$R(x) = \text{Remainder}\left[\frac{x^3 D(x)}{G_{CRC}(x)}\right]$$

In the case of error-free transmission, the code word $C'(X) = x^3 D(x) + R(x)$ is therefore divisible by $G_{CRC} C(x)$ without remainder. But since the check sum bits $R(x)$ are transmitted in inverted form, the division yields a remainder:

$$S(x) = \text{Remainder}\left[\frac{C(x)}{G_{CRC}(x)}\right] = \text{Remainder}\left[\frac{x^3 D(x) + \overline{R}(x)}{G_{CRC}(x)}\right] = x^2 + x + 1$$

This is equivalent to shifting the whole codeword $C(x)$ through an identical shift register on the decoder side, after priming it with C_{50}, C_{51}, C_{52}. After shifting in the last check sum bit (50 shift operations), this register should contain a 1. If this is not the case, the block contains faulty bits. Inversion of the parity bits avoids the generation of null code words, i.e. bursts which contain only zeros cannot occur on the traffic channel.

The speech data $d(k)$ ($k= 1, ..., 189$) of Class I of a block are combined with the parity bits $p(k)$ ($k=1, 2, 3$) and fill bits to form a new block $u(k)$ ($k=1, ..., 189$):

$$u(k) = \begin{cases} d(2k) & k = 0, ... , 90 \\ d(2 \times (184 - k) + 1) & k = 94, ... , 184 \\ p((k - 91) + 1) & k = 91, 92, 93 \\ 0 & k = 185, ... , 189 \end{cases}$$

The bits in even or odd positions are shifted to the upper or lower half of the block, respectively, and separated by the three check bits; additionally, the order of the odd bits is reversed. Finally the block is filled to 189 bits. Combination with the speech bits of Class II yields a block of 267 bits, which serves as input to the convolutional coder.

This enormous effort is taken because of the high compression rate and sensitivity against bit errors of the speech data. A speech frame in which the bits of Class I have been recognized as erroneous can therefore be reported as erroneous to the speech codec using the *Bad Frame Indication* (BFI); see Section 6.1. In order to maintain a constantly good speech quality, speech frames recognized as faulty are discarded, and the last correctly received frame is repeated, or an extrapolation of received speech data is performed.

6.2.2.2 Block Coding for Data Traffic Channels

Block coding of traffic channels is somewhat simpler for data services. In this case, no parity bits are determined. Blocks of length N0 arriving at the input of the coder are supplemented by fill bits to a size of N1 suitable for further coding. Table 6.3 gives an overview of the different block lengths, which depend on the data rate and channel type, i.e. whether the channel is a full-rate (TCH/Fxx) or half-rate (TCH/Hxx) channel.

The 9.6 kbit/s data service is only offered on a full-rate traffic channel. The data is offered in blocks of 60 bits to the channel coder. Four blocks each are combined and supplemented by four appended tail bits (zero bits). In the case of nontransparent data service, these four blocks make up exactly one protocol frame of the RLP protocol (240 bits).

The procedures for other data services are similar. Depending on whether they are offered over a half-rate or full-rate channel (see Table 5.1 and Table 6.3), blocks of 60 or 36 bit length (4.8 or $\leq$ 2.4 kbit/s) are constructed and supplemented with tail bits (zero bits) to form blocks of 76 or 244 bits respectively.

Table 6.3: Block formation for data traffic channels

Data channel	N0	Tail bits	N1
TCH/F9.6	4 $\times$ 60	4	244
TCH/F4.8	2 $\times$ 60	2 $\times$ 16	2 $\times$ 76
TCH/F2.4	2 $\times$ 36	4	76
TCH/H4.8	4 $\times$ 60	4	244
TCH/H2.4	4 $\times$ 36	2 $\times$ 4	2 $\times$ 76

6.2.2.3 Block Coding for Signaling Channels

The majority of the signaling channels (SACCH, FACCH, SDCCH, BCCH, PCH, AGCH) use an extremely powerful block code for error detection. This is a so-called fire code, i.e. a shortened binary cyclic code which appends 40 redundancy bits to the 184-bit data block. Its pure error detection capability is sufficient to let undetected errors go through only with a probability of 2^{-40}. (A FIRE code can also be used for error correction, but here it is used only for error detection.) Error detection with the fire code in the SACCH channel is used to verify connectivity (Figure 5.23), and is used, if indicated, to decide about breaking a connection. The fire code can be defined like the CRC by way of a generating polynomial:

$$TG_F(x) = (x^{23} + 1)(x^{17} + x^3 + 1)$$

The check sum bits $R_F(x)$ of this code are calculated in such a way that a 40-bit remainder $S_F(x)$ is left after dividing the code word $C_F(x)$ by the generating polynomial $G_F(x)$. In the case of no errors, the remainder contains only "1" bits:

$$S_F(x) = \text{Remainder}\left[\frac{C_F(x)}{G_F(x)}\right] = \text{Remainder}\left[\frac{x^{40} D_F(x) + R_F(x)}{G_F(x)}\right]$$

$$= x^{39} + x^{38} + \dots + x^2 + x + 1$$

The code word generated with the redundancy bits of the fire code is supplemented with "0" bits to a total length of 228 bits, which are then delivered to the convolutional coder.

Another approach has been used for error detection in the RACH channel. The very short random access burst in the RACH allows only a data block length of P0 = 8 bits, which is supplemented in a cyclic code by six redundancy bits . The corresponding generating polynomial is

$$G_{RACH}(x) = x^6 + x^5 + x^3 + x^2 + x + 1$$

In the *Access Burst* (AB), the mobile station also has to indicate a target base station. The BSIC of the respective base station is used for this purpose. The six bits of the BSIC are added to the six redundancy bits modulo 2, and the resulting sequence is inserted as the redundancy of the data block. The total code word to be convolution-coded for the RACH thus has a length of 18 bits; i.e. four fill bits (0) are also added in the RACH to this block. In exactly the same way, block coding is performed for the *Handover Access* burst, which is in principle also a random access burst.

The SCH channel, as an important synchronization channel, uses a somewhat more elaborate error protection than the RACH channel. The SCH data blocks have a length of 25 bits and receive, besides the fill bits, another 10 bits of redundancy for error detection through a cyclic code with somewhat better error detection capability than on the RACH:

$$G_{SCH}(x) = x^{10} + x^8 + x^6 + x^5 + x^4 + x^2 + 1$$

Thus the length of the code words delivered to the channel coder in the SCH channel is 39 bits. Table 6.4 summarizes the block parameters of the RACH and SCH channels. Table 6.5 presents an overview of the cyclic codes used in GSM.

Table 6.4: Block lengths for the RACH and SCH channels

Data channel	P0	Redundancy bits	Tail bits	P1
RACH	8	6	4	18
SCH	25	10	4	39

Table 6.5: Cyclic codes used for block coding in GSM

Channel	Polynomial
TCH/FS	$x^3 + x + 1$
DCCH and CCCH (part.)	$(x^{23} + 1)(x^{17} + x^3 + 1)$
RACH	$x^6 + x^5 + x^3 + x^2 + x + 1$
SCH	$x^{10} + x^8 + x^6 + x^5 + x^4 + x^2 + x + 1$

6.2.3 Internal Error Protection: Convolutional Coding

After block coding has supplemented the data with redundancy bits for error detection (parity bits), added fill bits and thus generated sorted blocks, the next stage is calculation of additional redundancy for error correction to correct the transmission errors caused by the radio channel. The internal error correction of GSM is based exclusively on convolutional codes.

Convolutional codes can also be defined like a CRC using shift registers and generating polynomials. Figure 6.8 shows the principle of convolutional coding. It consists in essence of using a shift register with K storage locations ($K = 5$ in Figure 6.8) where K is also known as the constraint length of the convolutional code. One data symbol d_i is read into the shift register per tact interval. A data symbol consists of k (here $k = 1$) data bits, each of which is moved into a storage location of the shift register. Theoretically a data symbol could consist of more than one bit ($k > 1$), but this is not realized in GSM. The data symbol read is combined with up to $(K-1)$ of its predecessor symbols $d_{i-1}, ..., d_{i-(K-1)}$ in several modulo 2 additions. The results of these operations are given to the interleaver as coded user payload symbols c_j. The value $(K-1)$ determines the number of predecessor symbols to be combined with a data symbol and is therefore also called the "memory" of the convolutional coder. The number v of combinatorial rules ($v=2$ in Figure 6.8) determines the number of coded bits in a code symbol c_j generated for each input symbol d_i. In Figure 6.8, the combinatorial results are scanned from top to bottom to generate the code symbol c_j. The combinatorial rules are defined by the generating polynomial $G_i(d)$.

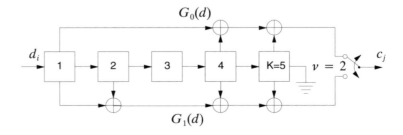

Figure 6.8: Principle of a convolutional coder

The rate r of a convolutional coder indicates how many data bits are processed for each coded bit, or $1/r$ is the number of coded bits per data bit. This rate is the essential measure of the redundancy produced by the code and hence its error correction capability:

$$r = k/v; \quad \text{in GSM:} \quad r_{GSM} = 1/v_{GSM} = \frac{1}{2}$$

The code rate of a convolutional coder is therefore determined by the number of bits k per input data symbol and the number of combinatorial rules v which are used for the calculation of a code symbol. In combination with the memory $(K-1)$, the code rate r determines the error correction capability of the code. In a simplified way: with decreasing r and increasing K the number of corrigible errors per code word increases, the error correction capabilities of the code are improved.

The coding procedure proper is expressed in the combinatorial operations (modulo 2 additions). These coding rules can be described with polynomials. In the case of the convolutional coder of Figure 6.8, the two generating polynomials are

$$G_0(d) = d^4 + d^3 + 1$$
$$G_1(d) = d^4 + d^3 + d + 1$$

Giving the generating polynomials makes a compact representation of the coding procedure possible. The maximal exponent indicates the memory $(K-1)$ of the convolutional coder, whereas the number of polynomials gives the rate r.

GSM defines different convolutional codes for the different logical channels (Figure 6.9). All have the same constraint length $K=5$, but they differ in the code rate and the polynomials used.

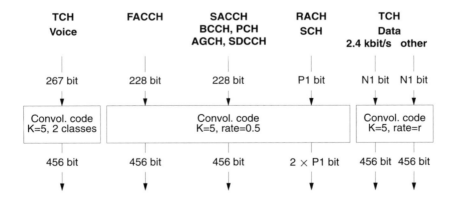

Figure 6.9: Overview of convolutional coding of logical channels

The coding procedures for the different logical channels are contained in four polynomials. Table 6.6 lists these polynomials G0, G1, G2, and G3. They are used in different combinations for convolutional coding of logical channels.

Table 6.6: Generating polynomials for convolutional codes

Type	Polynomial
G0	$1 + d^3 + d^4$
G1	$1 + d + d^3 + d^4$
G2	$1 + d^2 + d^4$
G3	$1 + d + d^2 + d^3 + d^4$

Table 6.7 give an overview of the uses and combinations of generating polynomials. It is evident that most logical channels use a convolutional code of rate 1/2 based on polynomials G0 and G1. Only the 4.8 kbit/s data service on a full-rate channel and the 2.4 kbit/s data service on a half-rate channel use a code of rate 1/3 based on polynomials G1, G2, and G3. The 2.4 kbit/s data service on a full-rate channel uses these three polynomials twice in a row to generate a convolutional code of rate 1/6. Convolutional coding of Class I speech bits on the speech channel generates a block of $2 \times 189 + 78 = 456$ bits. This is the uniform block size needed for mapping these data blocks onto the bursts with 114-bit payload. In a similar way, for most of the remaining channels, two coded blocks of 228 bits are combined.

Table 6.7: Usage of generating polynomials

Channel type	Generating polynomial			
	G0	G1	G2	G3
TCH, full-rate				
Voice Class I	■	■		
Voice Class II				
TCH, full-rate, 9.6 kbit/s	■	■		
TCH, full-rate, 4.8 kbit/s		■	■	■
TCH, half-rate, 4.8 kbit/s	■	■		
TCH, full-rate, 2.4 kbit/s		■	■	■
TCH, half-rate, 2.4 kbit/s		■	■	■
FACCHs	■	■		
SDCCHs, SACCHs	■	■		
BCCH, AGCH, PCH	■	■		
RACH	■	■		
SCH	■	■		

The situation for the data services TCH/F9.6 and TCH/H4.8 is somewhat different. At the input of the convolution coder, blocks of 244 bits arrive which the coder converts to blocks of 488 bits. These blocks are reduced to 456 bits by removing every 15th bit beginning with the 11th bit, i.e. a total of 32 bits are removed. This procedure is known as *puncturing,* and the resulting code is a punctured convolutional code [34][35][36]. On one hand, punctu-ring cuts the block size to a length suitable for further processing; on the other hand, punc-turing removes redundancy. The resulting net code rate of $r' = 244/456$ is therefore some-what higher than the rate of 1/2 for the convolution coder. Thus the code has slightly lower error correction capability. Puncturing cuts down the convolutional coded blocks of the TCH/F9.6 and TCH/H4.8 channels to the standard format of 456 bits. Thus blocks of these channels can also be processed in a standardized way (interleaving, etc.), and the amount of redundancy contained in a block is also matched to the bit rate available for transmis-sion.

Block coding appends at least four zero bits to each block (see Section 6.2.2). These bits not only serve as fill bits at the end of a block, but they are also important for the channel coding procedure. Shifted at the end of each block into the coder, these bits serve to reset the coder into the defined starting position, such that in principle adjacent data blocks can be coded independently of each other.

In most cases, the decoding of convolutional code employs the Viterbi algorithm. It uses a suitable metric to determine the data sequence that most likely equals the transmitted data (*maximum likelihood decoding*) [31]. Using the knowledge of the generating polyno-mials, the decoder can determine the original data sequence.

6.2.4 Interleaving

The decoding result of the convolutional coder is very dependent on the frequency and grouping of bit errors that occurr during transmission. Especially negative for error correc-tion are burst errors during longer and deeper fading periods, i.e. series of erroneous se-quential bits. In such cases, the channel is not a binary channel without memory, rather the single-bit errors have statistical dependence which diminishes the result of the error correction procedure of the convolutional coder. To achieve good error correction results, the channel should have no memory, i.e. the bit errors should be statistically independent. Therefore, burst errors occurring frequently on the radio channel should be distributed uniformly across the transmitted code words. This can be accomplished through the inter-leaving technique described in the following.

The interleaving approach is to distribute code words from the convolutional coder by spreading in time and merging them across several bursts for transmission. This principle is shown in Figure 6.10. By time spreading, each of the code words is distributed across a threefold length. Merging the bit sequences generated in this way has the effect that the individual bits from each of the three code words are sorted into alternate bursts; this way each code word is transmitted as distributed over a total of three bursts, and two bits of a data block are never transmitted adjacent to each other.

This kind of interleaving is also known as diagonal interleaving. The number of bursts over which a code word is spread is called the *interleaving depth*; a spreading factor can be de-fined analogously. A burst error in a transmission spurt is therefore distributed uniformly over several subsequently transmitted code words because of the distribution of the data over several bursts. This generates bit error sequences which are less dependently distrib-uted in the data stream, hence it improves the success of the error correction process. Figure 6.11 shows an example.

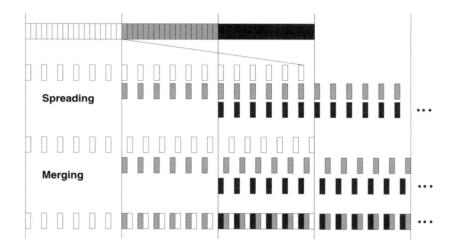

Figure 6.10: Interleaving: spreading and merging

During the third burst of transmission, severe fading of the signal leads to a massive burst error. This burst is now heavily affected by a total of six single-bit errors. In the process of deinterleaving (inversion of merging, despreading) these bit errors are distributed across three data blocks, corresponding to the bit positions which were sorted into the respective bursts during interleaving. The number of errors per data block is now only two, which can be much more easily corrected.

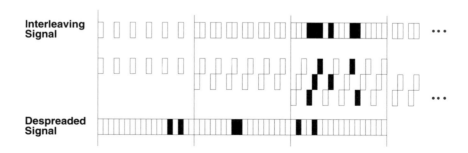

Figure 6.11: Distributing bit errors through deinterleaving

Another kind of interleaving is block interleaving. In this way, code words are written line by line into a matrix (Figure 6.12), which is subsequently read out column by column. The number of lines of the interleaving matrix determines the interleaving depth. As long as the length of a burst error is shorter than the interleaving depth, the burst error generates only single-bit errors per data block if block interleaving is used [29][31].

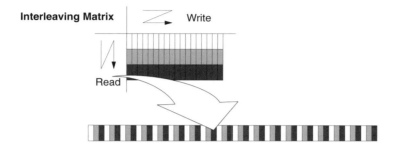

Figure 6.12: Principle of block interleaving

However, the great advantage of interleaving, to alleviate the effect of burst errors for optimal error correction with a convolutional coder, is traded for a not insignificant disadvantage for speech and data communication. As evident from Figure 6.10 and Figure 6.12,
the code words are spread across several bursts (here three). For a complete reconstruction of a code word, one has to wait for the complete transmission of three bursts. This
forces a transmission delay, which is a function of the interleaving depth. In GSM, both
methods of interleaving are used (Figure 6.13), block interleaving as well as bit interleaving. With a maximal interleaving depth of 19, this can lead to delays of up to 360 ms
(Table 6.8).

Table 6.8: Transmission delay caused by interleaving

Channel type	Interleav. depth	Transmission delay [ms]
TCH, full-rate, voice	8	38
TCH, full-rate, 9.6 kbit/s	22 (19)	93
TCH, full-rate, 4.8 kbit/s	22 (19)	93
TCH, half-rate, 4.8 kbit/s	22 (19)	185
TCH, full-rate, 2.4 kbit/s	8	38
TCH, half-rate, 2.4 kbit/s	22 (19)	185
FACCH, full-rate	8	38
FACCH, half-rate	8	74
SDCCH	4	14
SACCH/TCH	4	360
SACCH/SDCCH	4	14
BCCH, AGCH, PCH	4	14

The speech channel TCH/FS in GSM uses block-diagonal interleaving. A data block of 456
coded bits is distributed over 8 blocks. In this way, a new data block is started after each
fourth merged block.

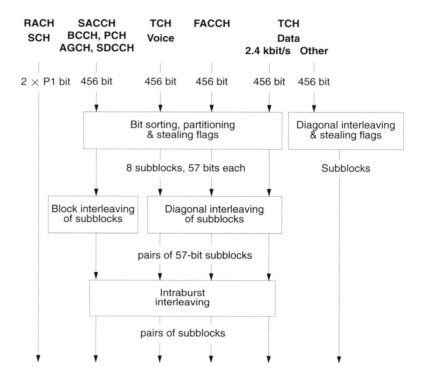

Figure 6.13: Overview: interleaving of logical channels

The exact interleaving rule for mapping the nth code word $c(n,k)$ with $k=456$ bits onto the bth interleaving block $i(b,j)$ with $j=114$ bits is

$$i(b,j) = c(n,k)$$

$$\text{with} \begin{cases} n = 0, 1, 2, \ldots, N, N+1, \ldots \\ k = 0, 1, 2, \ldots, 455 \\ b = b_0 + 4n + k \bmod 8 \\ j = 2 \times ((49\,k) \bmod 57) + ((k \bmod 8)\,\mathrm{div}\,4) \end{cases}$$

The data of the nth code word (data block n in Figure 6.14) are distributed across eight interleaving blocks at 114 bits each, beginning with block $B=b_0+4n$. In this way, only the even bits of the first four blocks $(B+0,1,2,3)$ and the odd bits of the last four blocks $(B+4,5,6,7)$ are used. The even bits of the last four blocks $(B+4,5,6,7)$ are occupied by data from block $n+1$. Each interleaving block thus contains bits of the current data block n and 57 bits of the following data block $n+1$ or the preceding block $n-1$, respectively.

The individual bits of data block n are alternatively distributed across the interleaving blocks, e.g. every eighth bit is in the same interleaving block according to the term $(k \bmod 8)$, whereas bit position j within an interleaving block $b=B+0, 1, 2, \ldots, 7$ is determined by two terms: the term $(k \bmod 8)$ div 4 is used to determine the even/odd bit positions; and the term $2 \times ((49 \times k) \bmod 57)$ determines the offset within the interleaving block. The first interleaving block B derived from data block n thus contains bits number 0, 8, 16, ..., 448, 456 of this data block.

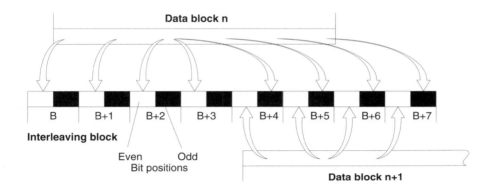

Figure 6.14: Interleaving TCH/FS: block mapping

The placement of these bits for the first block B in the interleaving block is illustrated in Figure 6.15. This placement is chosen in such a way that no two directly adjacent bits of the interleaving block belong to the same data block. In addition, the mapped bits are combined into groups of eight bits each, which are distributed as uniformly as possible across the entire interleaving block. This achieves additional spreading of error bursts within a data block. Therefore, the interleaving for the TCH/FS is block-diagonal interleaving with additional merging of data bits within the interleaving block. This is also called *intraburst interleaving* (Figure 6.13). The data channel TCH/F2.4 and the FACCH in GSM use the same interleaving methods as the TCH/FS.

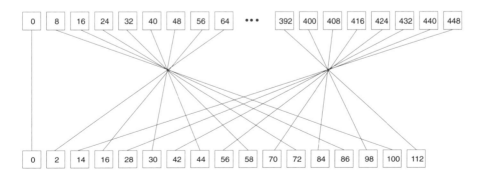

Figure 6.15: Mapping of data block n onto interleaving block B for a TCH/FS

For the other data services in the traffic channel (TCH/F9.6, TCH/F4.8, TCH/H4.8, TCH/H2.4) the interleaving is somewhat simpler. A pure bitwise diagonal interleaving with an interleaving depth of 19 is used. In this case, the interleaving rule is

$$i(b,j) \;=\; c(n,k)$$

$$\text{with} \begin{cases} n \;=\; 0, 1, 2, \ldots, N, N + 1, \ldots \\ k \;=\; 0, 1, 2, \ldots, 455 \\ b \;=\; b_0 + 4 \times n + k \bmod 19 + k \operatorname{div} 114 \\ j \;=\; k \bmod 19 \;+\; 19 \times (k \bmod 6) \end{cases}$$

The bits of a data block $c(n,k)$ are distributed in groups of 114 bits across 19 interleaving blocks, whereby groups of six bits are distributed uniformly over one interleaving block. With this diagonal interleaving, each interleaving block also starts a new 114-bit block of data. A closer look at this interleaving rule reveals that the input to the interleaver consists of blocks of 456 coded data bits as code words. The whole code word is therefore really spread across 22 interleaving blocks; the nominal interleaving depth of 19 results historically from 114-bit block interleaving.

Most signaling channels use an interleaving depth of 4, such as SACCH, BCCH, PCH, AGCH, and SDCCH. The interleaving scheme is almost identical to the one used for the TCH/FS, however, the code words $c(n,k)$ are spread across four rather than eight interleaving blocks:

$$i(b,j) \;=\; c(n,k)$$

$$\text{with} \begin{cases} n \;=\; 0, 1, 2, \ldots, N, N + 1, \ldots \\ k \;=\; 0, 1, 2, \ldots, 455 \\ b \;=\; b_0 + 4n + k \bmod 4 \\ j \;=\; 2 \times (\, (49\,k) \bmod 57 \,)) + (\, (k \bmod 8) \operatorname{div} 4 \,) \end{cases}$$

With this kind of interleaving, there are also eight blocks generated, just like in case of the TCH/FS, however, at the same time a block of 57 even bits is combined with a block of 57 odd bits to form a complete interleaving block. This has the consequence that consecutive coded signaling messages are not block-diagonally interleaved, but that each four consecutive interleaving blocks are fully occupied with the data of just one, and only one, code word. Also, a new code word starts after every four interleaving blocks. Therefore, this interleaving of GSM signaling messages is in essence also a block interleaving procedure. This is especially important for signaling channels to ensure the transmission of individual protocol messages independent of preceding or succeeding messages. This also enables some kind of asynchronous communication of signaling information. The signaling data of the RACH and SCH must each be transmitted in single data bursts; no interleaving occurs.

6.2.5 Mapping onto the Burst Plane

After block convolutional coding and interleaving, the data is available in form of 114-bit interleaving blocks. This corresponds exactly to the amount of data which can be carried by a normal burst (Figure 5.6). Each interleaving block is mapped directly onto one burst (Figure 6.16). After setting the stealing flags, the bursts can be composed and passed to the modulator. The stealing flags indicate whether high-priority signaling messages are present (FACCH messages), which must be transmitted as fast as possible, instead of the originally planned data of the traffic channel.

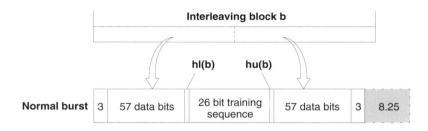

Figure 6.16: Mapping onto a burst

An essential component of GSM channel coding is the correct treatment of FACCH signaling messages which are multiplexed in a preemptive way into the traffic channel. At the burst level, each FACCH code word displaces a code word of the current TCH/FS traffic channel, i.e. the code words must be tied into the interleaving structure instead of regular data blocks of the traffic channel. The interleaving rule for the FACCH is the same as for the TCH/FS: from an FACCH code word the even positions are occupied in one set of four interleaving blocks, and the odd bit positions are occupied in another set of interleaving blocks; in addition, the bit positions within the interleaving blocks are shuffled (intraburst interleaving).

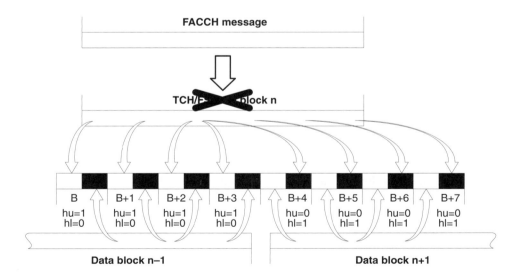

Figure 6.17: Insertion of an FACCH message into the TCH/FS data stream

When an FACCH message needs to be transmitted (e.g. a handover command), the current data block n is replaced by the convolutional coded FACCH message, and it is interleaved in a block-diagonal way with the data blocks $(n-1)$ and $(n+1)$ of the traffic channel

(Figure 6.17). In the eight blocks involved in this procedure, B, B+1, ..., B+7, the respective stealing flags hl(b) and hu(b) have to be set (Figure 6.16). If neither flag is set, the burst contains data of the traffic channel. If the even bits of the burst are occupied by FACCH data, hu(b) is set; in the case of the odd bits being used for FACCH, hl(b) is set (Figure 6.17).

If the current burst is not available for traffic channel data, the data block n has to be discarded. Bits "stolen" in this way have varying effects:

- A complete speech frame of the TCH/FS is lost (20 ms speech).
- In the case of TCH/F9.6 and TCH/H4.8 channels, three bits are stolen from each of the eight interleaving blocks, which belong to the same data block, such that a maximum of 24 coded data bits are interfered with.
- In the case of TCH/F4.8 and TCH/H2.4 channels, 6 bits are stolen from each of the eight interleaving blocks, which belong to the same data block, such that a maximum of 48 coded data bits are interfered with.
- In the case of TCH/F2.4 channels, the same interleaving rules as in the TCH/FS are used, such that a complete data block is displaced by the FACCH.

In summary, the FACCH signaling needed for fast reactions causes data losses or bit errors in the accompanying traffic channel, and they have to be totally or partially corrected by the convolutional coder.

6.3 Security-Related Network Functions and Encryption

Methods of encryption for user data and for the authentication of subscribers, like all techniques for data security and data protection, are gaining enormous importance in modern digital systems [3]. GSM therefore introduced powerful algorithms and encryption techniques. The various services and functions concerned with security in a GSM PLMN are categorized in the following way:

- Subscriber Identity Confidentiality
- Subscriber Identity Authentication
- Signaling Information Element Confidentiality
- Data Confidentiality for Physical Connections

In the following section, the security functions concerning the subscriber are presented.

6.3.1 Protection of Subscriber Identity

The intent of this function is to prevent disclosing which subscriber is using which resources in the network, by listening to the signaling traffic on the radio channel. On one hand this should ensure the confidentiality of user data and signaling traffic, on the other hand it should also prevent localizing and tracking of a mobile station. This means above all that the *International Mobile Subscriber Identity* (IMSI) should not be transmitted as clear text, i.e. unencrypted.

Instead of the IMSI, one uses a *Temporary Mobile Subscriber Identity* (TMSI) on the radio channel for identification of subscribers. The TMSI is temporary and has only local validity, which means that a subscriber can only be uniquely identified by TMSI and the *Location Area ID* (LAI). The association between IMSI and TMSI is stored in the VLR.

The TMSI is issued by the VLR, at the latest, when the mobile station changes from one Location Area (LA) into another (location updating). When a new location area is entered, this is noticed by the mobile station (Section 3.2.5) who reports to the new VLR with the old LAI and TMSI (LAI*old* and TMSI*old*, Figure 6.18). The VLR then issues a new TMSI for the MS. This TMSI is transmitted in encrypted form.

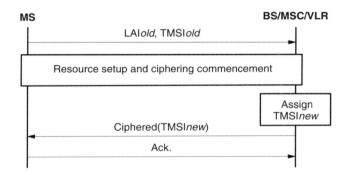

Figure 6.18: Encrypted transmission of the temporary subscriber identity

The subscriber identity is thus protected against eavesdropping in two ways: first, the temporary TMSI is used on the radio channel instead of the IMSI; second, each new TMSI is transmitted in encrypted form.

In the case of database failures, if the VLR database is partially lost or no correct subscriber data is available (loss of TMSI, TMSI unknown at VLR, etc.), the GSM standard provides for a positive acknowledgment of the subscriber identity. For this subscriber identification, the IMSI must be transmitted as clear text (Figure 6.19) before encryption is turned on. Once the IMSI is known, encryption can be restarted and a new TMSI can be assigned.

6.3.2   Verification of Subscriber Identity

When a subscriber is added to a home network for the first time, a *Subscriber Authentication Key* (Ki) is assigned in addition to the IMSI to enable the verification of the subscriber identity (also known as authentication). At the network side, the key Ki is stored in the authentication center (AUC) of the home PLMN. At the subscriber side, this key must be stored on the SIM card of the subscriber.

The process of authenticating a subscriber is essentially based on the A3 algorithm, which is performed at the network side as well as at the subscriber side (Figure 6.20). This algorithm calculates independently on both sides (MS and network) the *Signature Response* (SRES) from the authentication key Ki and a *Random Number* (RAND) offered by the

network. The MS transmits its SRES value to the network which compares it with its calculated value. If both values agree, the authentication was successful. Each execution of the
algorithm A3 is performed with a new value of the random number RAND which cannot
be predetermined; in this way recording the channel transmission and playing it back cannot be used to fake an identity.

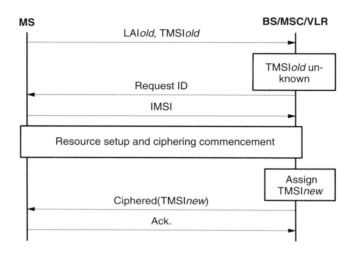

Figure 6.19: Clear text transmission of the IMSI when the TMSI is unknown

6.3.3 Generating Security Data

At the network side, the 2-tuple (RAND, SRES) need not be calculated each time when
authentication has to be done. Rather the AUC can calculate a set of (RAND, SRES) 2-tu-
ples in advance, store them in the HLR, and send them on demand to the requesting VLR.
The VLR stores this set (RAND[n], SRES[n]) and uses a new 2-tuple from this set for each
authentication procedure. Each 2-tuple is used only once; so new 2-tuples continue to be
requested from the HLR/AUC.

Figure 6.20: Principle of subscriber authentication

This procedure, to let security data (Kc, RAND, SRES) be calculated in advance by the AUC has the advantage that the secret authentication key Ki of a subscriber can be kept exclusively within the AUC, which ensures a higher level of confidentiality. A somewhat less secure variant is to supply the currently needed key Ki to the local VLR which then generates the security data locally.

If the key Ki is kept exclusively in the AUC, the AUC has to generate a set of security data for a specific IMSI on demand from the HLR (Figure 6.21): the random number RAND is generated and the pertinent signature SRES is calculated with the A3 algorithm, whereas the A8 algorithm generates the encryption key Kc.

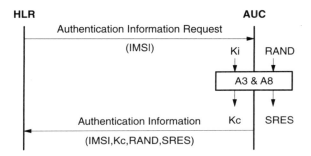

Figure 6.21: Generation of a set of security data for the HLR

The set of security data, a 3-tuple consisting of Kc, RAND, and SRES, is sent to the HLR and stored there. In most cases, the HLR keeps a supply of security data (e.g. 5), which can then be transmitted to the local VLR, so that one does not have to wait for the AUC to generate and transmit a new key. When there is a change of LA into one belonging to a new VLR, the sets of security data can be passed on to the new VLR. This ensures that the subscriber identity IMSI is transmitted only once through the air, namely when no TMSI has yet been assigned (see registration) or when this data has been lost. Afterwards the (encrypted) TMSI can be used for communicating with the MS.

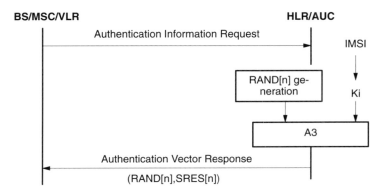

Figure 6.22: Highly secure authentication (no transmission of Ki)

If the IMSI is stored on the network side only in the AUC, all authentication procedures can be performed with the 2-tuples (RAND, SRES) which were precalculated by the AUC. Besides relieving the load on the VLR (no execution of the A3 algorithm), this kind of subscriber identification (Figure 6.22) has the other advantage of being particularly secure, because confidential data, especially Ki, need not be transmitted over the air. It should be used especially when the subscriber is roaming in a network of a foreign operator, since it avoids passing of security-critical data over the network boundary.

The less secure variant (Figure 6.23) should only be used within a PLMN. In this case, the secret (security-critical) key Ki is transmitted each time from the HLR/AUC to the current VLR, which executes the algorithm A3 for each authentication.

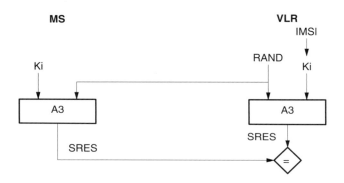

Figure 6.23: Weakly secure authentication (transmission of Ki to VLR)

6.3.4   Encryption of Signaling and Payload Data

The encryption of transmitted data is a special characteristic of GSM networks that distinguishes the offered service from analog cellular and fixed ISDN networks. This encryption is performed at the transmitting side after channel coding and interleaving and immediately preceding modulation (Figure 6.24). On the receiving side, decryption directly follows the demodulation of the data stream.

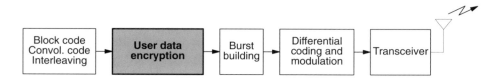

Figure 6.24: Encryption of payload data in the GSM transport chain

A *Cipher Key* (Kc) for the encryption of user data is generated at each side using the generator algorithm A8 and the random number RAND of the authentication process (Figure 6.25). This key Kc is then used in the encryption algorithm A5 for the symmetric encryption of user data. At the network side, the values of Kc are calculated in the AUC/ HLR simultaneously with the values for SRES. The keys Kc are combined with the 2-tuples (RAND, SRES) to produce 3-tuples, which are stored at the HLR/AUC and supplied on demand, in case the subscriber identification key Ki is only known to the HLR (Section 6.3.2). In the case of the VLR having access to the key Ki, the VLR can calculate Kc directly.

Figure 6.25: Generation of the cipher key Kc

The encryption of signaling and user data is performed at the mobile station MS as well as at the base station BTS (Figure 6.26). This is a case of symmetric encryption, i.e. ciphering and deciphering are performed with the same key Kc and the A5 algorithm.

Based on the secret key Ki stored in the network, the cipher key Kc for a connection or signaling transaction can be generated at both sides, and the BTS and MS can decipher each other's data. Signaling and user data are encrypted together (TCH/SACCH/ FACCH); for dedicated signaling channels (SDCCH) the same method is used as for traffic channels. This process is also called a *stream cipher*, i.e. ciphering uses a bit stream which is added bitwise to the data to be enciphered (Figure 6.27).

Figure 6.26: Principle of symmetric encryption of user data

Deciphering consists of performing an additional EXCLUSIVE OR operation of the enciphered data stream with the ciphering stream. The frame number (FN) of the current TDMA frame within a hyperframe (see Section 5.3.1) is another input for the A5 algorithm besides the key Kc, which is generated anew for each connection or transaction. The current frame number is broadcast on the synchronization channel (SCH) and is thus available any time to all mobile stations currently in the cell. Synchronization between ciphering and deciphering processes is thus performed through FN.

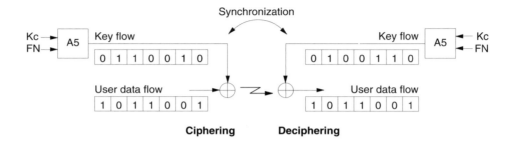

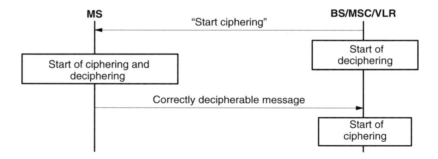

Figure 6.27: Combining payload data stream and ciphering stream

However, the problem of synchronizing the activation of the ciphering mode has to be solved first: the deciphering mechanism on one side has to be started at precisely the correct moment. This process is started under network control, immediately after the authentication procedure is complete or when the key Kc has been supplied to the base station (BS); see Figure 6.28.

The network, i.e. the BTS, transmits to the mobile station the request to start its (de)ciphering process, and it starts its own deciphering process. The mobile station then starts its ciphering and deciphering. The first ciphered message from the MS which reaches the network and is correctly deciphered leads to the start of the ciphering process on the network side.

Figure 6.28: Synchronized start of the ciphering process

7 Protocol Architecture

7.1 Protocol Architecture Planes

The various physical aspects of radio transmission across the GSM air interface and the realization of physical and logical channels were explained in Chapter 5. According to the terminology of the OSI Reference Model, these logical channels are at the Service Access Point of Layer 1 (physical layer), where they are visible to the upper layers as transmission channels of the physical layer. The physical layer also includes the forward error correction and the encryption of user data.

The separation of logical channels into the two categories of control channels (signaling channels) and traffic channels (Table 5.1) corresponds to the distinction made in the ISDN Reference Model between user plane and control plane. Figure 7.1 shows a simplified reference model for the GSM *User–Network Interface* (UNI) Um, where the layer-transcending management plane is not elaborated in the following. In the user plane, protocols of the seven OSI layers are defined for the transport of data from a subscriber or a data terminal. User data is transmitted in GSM across the air interface over traffic channels TCH, which therefore belong to Layer 1 of the user plane (Figure 7.1).

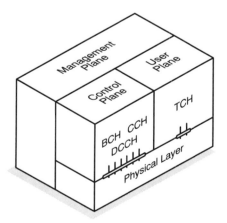

Figure 7.1: Logical channels at the air interface in the ISDN Reference Model

Protocols in the signaling plane are used to handle subscriber access to the network and for the control of the user plane (reservation, activation, routing, switching of channels and connections). In addition, signaling protocols between network nodes are needed (network internal signaling). The Dm channels of the air interface in GSM are signaling channels and are therefore realized in the signaling plane (Figure 7.1).

Since signaling channels are physically present but mostly unused during an active user connection, it is obvious to use them also for the transmission of certain user data. In ISDN, packet-switched data communication is therefore permitted on the D channel, i.e. the physical D channel carries multiplexed traffic of signaling data (s-data) and user (pay-load) data (p-data). The same possibility also exists in GSM. Data transmission without allocation of a dedicated traffic channel is used for the *Short Message Service* (SMS) by us-ing free capacities on signaling channels. For this purpose, a separate SDCCH is allocated, or, if a traffic connection exists, the SMS protocol data units are multiplexed onto the signaling data stream of the SACCH (Figure 7.2).

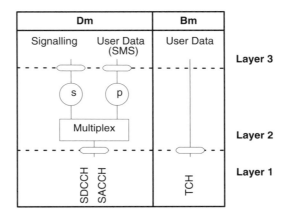

Figure 7.2: User data and control at the air interface

The control (signaling) and user plane can be defined and implemented separately of each other, ignoring for the moment that control and user data have to be transmitted across the same physical medium at the air interface and that signaling procedures initiate and control activities in the user plane. Therefore, for each plane there exists a corresponding separate protocol architecture within the GSM system: the user data protocol architecture (see Section 7.2) and the signaling protocol architecture (see Section 7.3), with an addi-tional separate protocol architecture for the transmission of p-data on the control (signal-ing) plane (see Section 7.3.2). A protocol architecture comprises not only the protocol en-tities at the radio interface Um but all protocol entities of the GSM network components.

7.2   Protocol Architecture of the User Plane

A GSM PLMN can be defined by a set of access interfaces (see Section 9.1) and a set of connection types used to realize the various communication services. A connection in GSM is defined between reference points. Connections are constructed from connection elements (Figure 7.3), and the signaling and transmission systems may change from element to element. Two elements therefore exist within a GSM connection: the radio interface connection element and the A interface connection element.

The radio interface and the pertinent connection element are defined between mobile station (MS) and base station subsystem (BSS), whereas the A interface connection element exists between BSS and MSC across the A interface. A GSM-specific signaling system is used at the radio interface, whereas ISDN-compatible signaling and payload transport are used across the A interface. The BSS is subdivided into BTS and BSC. Between them they define the Abis interface, which has no connection element defined; this is because it is usually transparent for user data.

Figure 7.3: Connection elements

A GSM connection type provides a way to describe GSM connections. Connection types represent the capabilities of the lower layers of the GSM PLMN. In the following section, the protocol models are presented as the basis for some of the connection types defined in the GSM standards. These are speech connections and transparent as well as nontransparent data connections. A detailed discussion of the individual connection types can be found in Chapter 9 with a description of how various data services have been realized in GSM.

7.2.1   Speech Transmission

The digital, source-coded speech signal of the mobile station is transmitted across the air interface in error-protected and encrypted form. The signal is then deciphered in the BTS, and the error protection is removed before the signal is passed on. This specially protected speech transmission occurs transparently between mobile station and a *Transcoding and Rate Adaptation Unit* (TRAU) which serves to transform the GSM speech-coded signals to the ISDN standard format (ITU-T A-law). A possible transport path for speech signals is shown in Figure 7.4, where the bit transport plane (encryption and TDMA/FDMA) has been omitted.

A simple GSM speech terminal (MT0, see also Figure 9.1) contains a *GSM Speech Codec* (GSC) for speech coding. Its speech signals are transmitted to the BTS after channel coding (FEC) and encryption, where they are again deciphered, decoded, and if necessary, error-corrected. More than one GSM speech signal can be multiplexed onto an ISDN channel, with up to four GSM speech signals (at 13 kbit/s each) per ISDN B channel (64 kbit/s). Before they are passed to the MSC, speech signals are transcoded in the BSS from GSM format to ISDN format (ITU-T A-law).

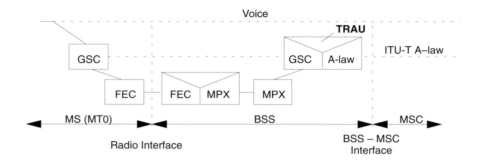

Figure 7.4: Speech transmission in GSM

The BTSs are connected to the BSC over digital fixed lines, usually leased lines or micro-wave links, with typical transmission rates of 2048 kbit/s (in Europe), 1544 kbit/s (in the USA) or 64 kbit/s (ITU-T G.703, G.705, G.732). For speech transmission, the BSS implements channels of 64 or 16 kbit/s. The physical placement of the *Transcoding and Rate Adaptation Unit* (TRAU) largely determines which kind of speech channel is used in the fixed network. The TRAU performs the conversion of speech data between GSM format (13 kbit/s) and ISDN A-law format (64 kbit/s). In addition, it is responsible for the adaptation of data rates, if necessary, for data services. There are two alternatives for the positioning of the TRAU: the TRAU can be placed into the BTS or outside of the BTS into the BSC. An advantage of placing the TRAU outside of the BTS is that up to four speech signals can be submultiplexed (MPX in Figure 7.4) onto an ISDN B channel, so that less bandwidth is required on the BTS-to-BSC connection. Beyond this consideration, placing the TRAU outside of the BTS allows the TRAU functions to be combined for all BTSs of a BSS in one separate hardware unit, perhaps produced by a separate manufacturer. The TRAU is, however, always considered as part of the BSS and not as an independent network element.

Figure 7.5 shows some variants of TRAU placement. A BTS consists of a *Base Control Function* (BCF) for general control functions like frequency hopping, and several (at least one) *Transceiver Function* (TRX) modules which realize the eight physical TDMA channels on each frequency carrier. The TRX modules are also responsible for channel coding and decoding as well as encryption of speech and data signals. If the TRAU is integrated into the BTS, speech transcoding between GSM and ISDN formats is also done within the BTS.

In the first case, TRAU within the BTS (BTS 1,2,3 in Figure 7.5), the speech signal in the BTS is transcoded into a 64 kbit/s A-law signal, and a single speech signal per B channel (64 kbit/s) is transmitted to the BSC/MSC. For data signals, the bit rates are adapted to 64 kbit/s, or several data channels are submultiplexed over one ISDN channel. The resulting user plane protocol architecture for speech transport is shown in Figure 7.6.

GSM-coded speech (13 kbit/s) is transmitted over the radio interface (Um) in a format that is coded for error protection and encryption. At the BTS site, the GSM signal is transcoded into an ISDN speech signal and transmitted transparently through the ISDN access network of the MSC.

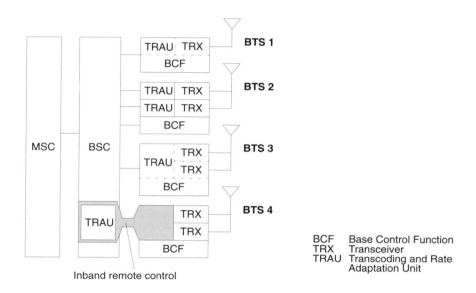

Figure 7.5: BTS architecture variations and TRAU placement

In the second case, the TRAU resides outside of the BTS (BTS 4 in Figure 7.5) and is considered a part of the BSC. However, physically it could also be located at the MSC site, i.e. at the MSC side of the BSC-to-MSC links (Figure 7.7). Channel coding/decoding and encryption are still performed in the TRX module of the BTS, whereas speech transcoding takes place in the BSC. For control purposes, the TRAU needs to receive synchronization and decoding information from the BTS, e.g. *Bad Frame Indication* (BFI) for error concealment (see Section 6.1). If the TRAU does not reside in the BTS, it must be remotely controlled from the BTS by inband signaling. For this purpose, a subchannel of 16 kbit/s is reserved for the GSM speech signal on the BTS-to-BSC link, so an additional 3 kbit/s is made available for inband signaling. Alternatively, the GSM speech signal with added inband signaling could also be transmitted in a full ISDN B channel.

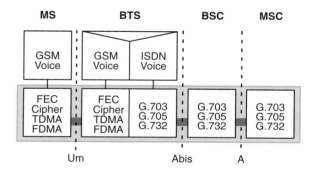

Figure 7.6: GSM protocol architecture for speech (TRAU at BTS site)

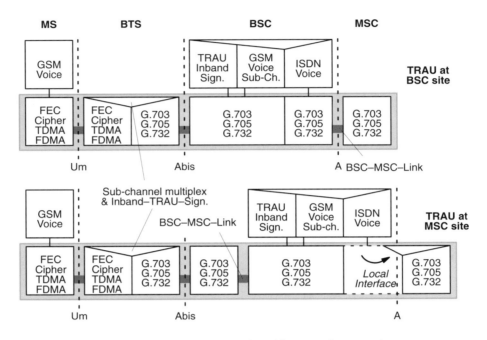

Figure 7.7: GSM protocol architecture for speech

7.2.2 Transparent Data Transmission

The digital mobile radio channel is subject to severe quality variations and generates burst errors, which one tries to correct through interleaving and convolutional codes (see Section 6.2). However, if the signal quality is too low due to fading breaks or interference, the resulting errors cannot be corrected. For data transmission across the air interface Um, a residual bit error ratio varying between 10^{-2} and 10^{-5} according to channel conditions can be observed [4]. This kind of variable quality of data transmission at the air interface determines the service quality of transparent data transmission. Transparent data transmission defines a GSM connection type used for the realization of some basic bearer services (transparent asynchronous and synchronous data, Table 4.2). The pertinent protocol architecture is illustrated in Figure 7.8. The main aspect of the transparent connection type is that user data is protected against transmission errors by forward error correction only across the air interface. Further transmission within the GSM network to the next MSC with an interworking function (IWF) to an ISDN or a PSTN occurs unprotected on digital line segments, which have anyway a very low bit error ratio in comparison to the radio channel. The transparent GSM data service offers a constant throughput rate and constant transmission delay; however, the residual error ratio varies with channel quality due to the limited correction capabilities of the FEC.

For example, take a data terminal communicating over a serial interface of type V.24. A transparent bearer service provides access to the GSM network directly at a mobile station or through a terminal adapter (reference point R in Figure 9.1). A data rate of up to 9600 bit/s can be offered based on the transmission capacity of the air interface and using an appropriate bit rate adaptation. The bit rate adaptation also performs the required asynchronous-to-synchronous conversion at the same time. This involves supplementing the tokens arriving asynchronously from the serial interface with fill data, since the channel coder requires a fixed block rate. This way there is a digital synchronous circuit-switched connection between the terminal accessing the service and the IWF in the MSC, which extends across the air interface and the digital ISDN B channel inside the GSM network; this synchronous connection is completely transparent for the asynchronous user data of the terminal equipment (TE).

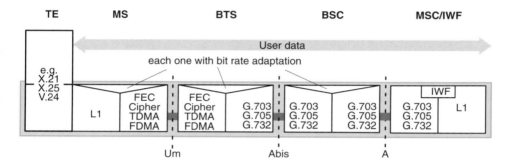

Figure 7.8: GSM protocol architecture for transparent data

7.2.3 Nontransparent Data Transmission

Compared to the bit error ratio of the fixed network, which is on the order of 10^{-6} to 10^{-9}, the quality of transparent data service is often insufficient for many applications, especially under adverse conditions. To provide more protection against transmission errors, more redundancy has to be added to the data stream. Since this redundancy is not always required, but only when there are residual errors in the data stream, forward error correction is inappropriate. Rather, an error detection scheme with automatic retransmission of faulty blocks is used, *Automatic Repeat Request* (ARQ). Such an ARQ scheme which was specifically adapted to the GSM channel, is the *Radio Link Protocol* (RLP). The assumption for RLP is that the underlying forward error correction of the convolutional coder realizes a channel with an average block error ratio of less than 10%, with a block corresponding to an RLP protocol frame of length 240 bits. Now the nontransparent channel experiences a constantly lower bit error ratio than the transparent channel, independent of the varying transmission quality of the radio channel; however, due to the RLP-ARQ procedure both throughput and transmission delay vary with the radio channel quality.

The data transmission between mobile station and interworking function of the next MSC is protected with the data link layer protocol RLP, i.e. the endpoints of RLP terminate in MS and IWF entities, respectively (Figure 7.9). At the interface to the data terminal TE, a *Nontransparent Protocol* (NTP) and an *Interface Protocol* (IFP) are defined, depending on the nature of the data terminal interface. Typically, a V.24 interface is used to carry character-oriented user data. These characters of the NTP are buffered and combined into blocks in the *Layer 2 Relay* (L2R) protocol, which transmits them as RLP frames. The data transport to and from the data terminal is flow-controlled. Therefore, transmission within the PLMN is no longer transparent for the data terminal. At the air interface, a new RLP frame is transmitted every 20 ms; thus L2R may have to insert fill tokens, if a frame cannot be completely filled at transmission time.

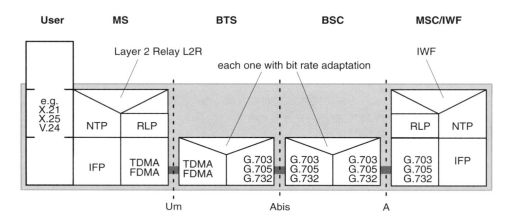

Figure 7.9: GSM protocol architecture for nontransparent data

The RLP protocol is very similar to the HDLC of ISDN with regard to frame structure and protocol procedures, the main difference being the fixed frame length of 240 bits, in contrast to the variable length of HDLC. The frame consists of a 16-bit protocol header, 200-bit information field, and a 24-bit *Frame Check Sequence* (FCS); see Figure 7.10. Because of the fixed frame length, the RLP has no reserved flag pattern, and a special procedure to realize code transparency like bit stuffing in HDLC is not needed. The very short – and hence less error prone – frames are exactly aligned with channel coding blocks. (The probability of frame errors increases with the length of the frame.)

RLP makes use of the services of the lower layers to transport its protocol data units (PDUs). The channel offered to RLP therefore has the main characteristic of a 200 ms transmission delay, besides the possibly occurring residual bit errors. The delay is mostly caused by interleaving and channel coding, since the transmission itself takes only about 25 ms for a data rate of 9600 bit/s. This means it will take at least 400 ms until a positive acknowledgment is received for an RLP frame, and protocol parameters like transmission window and repeat timers need to be adjusted accordingly.

The RLP header is similar to the one used in HDLC [15], with the difference that the RLP header contains no address information but only control information for which 16 bits are available. One distinguishes *supervisory frames* and *information frames*. Whereas information frames carry user data, supervisory frames serve to control the connection (initialize, disconnect, reset) as well as the retransmission of information frames during data transfer. The information frames are labeled with a sequence number N(S) for identification, for which 6 bits are available in the RLP header (Figure 7.10). To conserve space, this field is also used to code the frame type. Sequence number values smaller than 62 indicate that the frame carries user data in the information field (information frame). Otherwise the information field is discarded, and only the control information in the header is of interest (supervisory frame). These frames are marked with the reserved values 62 and 63 (Figure 7.10).

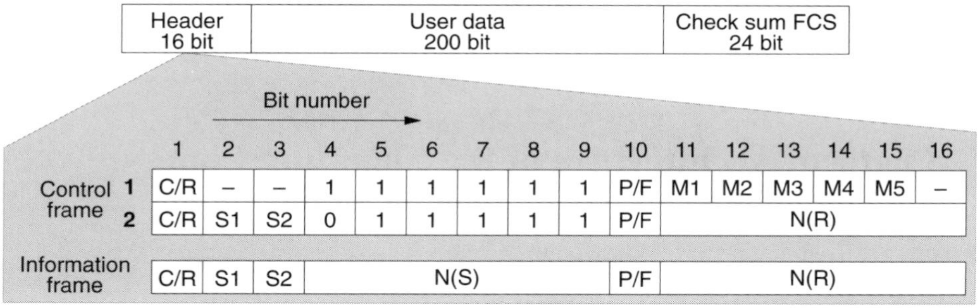

Figure 7.10: Frame structure of the RLP protocol

Due to this header format, information frames can also carry (implicit) control information, a process known as piggybacking. The header information of the second variant can be carried completely within the header of an information frame. This illustrates further how RLP has been adapted to the radio channel, since it makes the transmission of additional control frames unnecessary during information transfer, which reduces the protocol overhead and increases the throughput.

Thus the send sequence number is calculated modulo 62, which amounts to a window of 61 frames, allowing 61 outstanding frames without acknowledgment before the sender has to receive the acknowledgment of the first frame. Positive acknowledgment is used; i.e. the receiver sends an explicit supervisory frame as a receipt or an implicit receipt within an information frame. Such an acknowledgment frame contains a receive frame number $N(R)$ which designates correct reception of all frames, including send sequence number $N(S) = N(R) - 1$.

Each time the last information frame is sent, a timer T1 is started at the sender. If an acknowledgment for some or all sent frames is not received in time, perhaps because the acknowledging RLP frame had errors and was therefore discarded, the timer expires and causes the sender to request an explicit acknowledgment. Such a request may be repeated N2 times; if this still leads to no acknowledgment, the connection is terminated. If an acknowledgment $N(R)$ is obtained after expiration of timer T1, all sent frames starting from and including $N(R)$ are retransmitted. In the case of an explicitly requested acknowledgment, this corresponds to a modified *Go-back-N* procedure. Such a retransmission is also allowed only up to N2 times. If no receipt can be obtained even after N2 trials, the RLP connection is reset or terminated.

Two procedures are provided in RLP for dealing with faulty frames: *selective reject*, which selects a single information frame without acknowledgment; and *reject*, which causes retransmission with implicit acknowledgment. With *selective reject*, the receiving RLP entity requests retransmission of a faulty frame with sequence number $N(R)$, but this does not acknowledge receipt of other frames. Each RLP implementation must at least include the *reject* method for requesting retransmission of faulty frames. With a *reject*, the receiver asks for retransmission of all frames starting with the first defective received frame with number $N(R)$ (*Go-back-N*). Simultaneously, this implicitly acknowledges correct reception of all frames up to and including $N(R)-1$. Realization of *selective reject* is not mandatory in RLP implementations, but it is recommended. The reason is that *Go-back-N* causes retransmission of frames that may have been transmitted correctly and thus deteriorates the throughput that could be achieved with selective reject.

7.3   Protocol Architecture of the Signaling Plane

7.3.1   Overview of the Signaling Architecture

Figure 7.11 shows the essential protocol entities of the GSM signaling architecture (control plane or signaling plane). Three connection elements are distinguished: the radio-interface connection element, the BSS-interface connection element, and the A-interface connection element. This control plane protocol architecture consists of a GSM-specific part with the interfaces Um and Abis and a part based on *Signaling System Number 7* (SS#7) with the interfaces A, B, C, E (Figure 7.11). This change of signaling system corresponds to the change from radio interface connection element to A-interface connection element as discussed above for the user data plane (Figure 7.3).

The radio interface Um is defined between mobile station and base station subsystem BSS, more exactly between MS and BTS. Within the BSS, the BTS and the BSC cooperate over the Abis interface, whereas the A interface is located between BSC and MSC. The MSC has also signaling interfaces to VLR (B), HLR (C), to other MSCs (E), and to the EIR (F). Further signaling interfaces are defined between VLRs (G) and between VLR and HLR (D). Figure 3.9 gives an overview of the interfaces in a GSM PLMN.

In the control plane, the lowest layer of the protocol model at the air interface, the *Physical Layer,* implements the logical signaling channels (TDMA/FDMA, multiframes, channel coding, etc.; see Chapter 5, Sections 6.1, 6.2, and 6.3). Like user data, signaling messages are transported over the Abis interface (BTS-BSC) and the A interface (BSC-MSC) on digital lines with data rates of 2048 kbit/s (1544 kbit/s in the USA), or 64 kb/s (ITU-T G.703, G.705, G.732).

On Layer 2 of the logical signaling channels across the air interface, a data link protocol entity is implemented, the *Link Access Procedure* on *Dm* channels (LAPDm). LAPDm is a derivative of LAPD which is specifically adapted to the air interface. This data link protocol is responsible for the protected transfer of signaling messages between MS and BTS over the air interface, i.e. LAPDm is terminated in mobile station and base station.

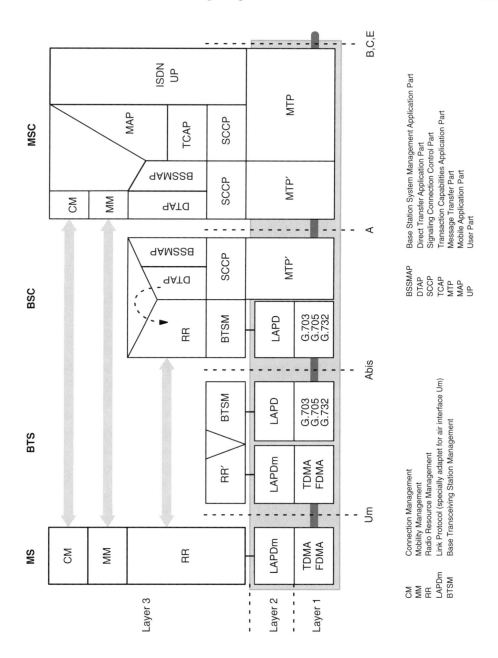

Figure 7.11: GSM protocol architecture for signaling

In essence, LAPDm is a protocol similar to HDLC which offers a number of services on the various logical Dm channels of Layer 3: connection setup and teardown, protected signaling data transfer. It is based on various link protocols used in fixed networks, such as LAPD in ISDN [22]. The main task of LAPDm is the transparent transport of messages between protocol entities of Layer 3 with special support for

- Multiple entities in Layer 3 and Layer 2
- Signaling for broadcasting (BCCH)
- Signaling for paging (PCH)
- Signaling for channel assignment (AGCH)
- Signaling on dedicated channels (SDCCH)

A detailed discussion of LAPDm will be presented in Section 7.4.2.

In the mobile station, the LAPDm services are used at Layer 3 of the signaling protocol architecture. There, Layer 3 is divided into three sublayers: *Radio Resource Management* (RR), *Mobility Management* (MM), and *Connection Management* (CM). The protocol architecture formed by these three sublayers is shown in Figure 7.12. *Connection management* is further subdivided into three protocol entities: *Call Control* (CC), *Supplementary Services* (SS), and *Short Message Service* (SMS). Additional multiplexing functions within Layer 3 are required between these sublayers.

The call-independent supplementary services and the short message service are offered to higher layers at two *Service Access Points* (SAPs), MNSS and MNSMS. A more detailed look at the services offered by the RR, MM, and CC protocol entities will be given in the following.

Radio resource management (RR) essentially handles the administration of the frequencies and channels. This involves the RR module of the MS communicating with the RR module of the BSC (Figure 7.11). The general objective of the RR is to set up, maintain, and take down RR connections which enable point-to-point communication between MS and network. This also includes *cell selection* in idle mode and handover procedures. Furthermore, the RR is responsible for monitoring BCCH and CCCH on the downlink when no RR connections are active.

The following functions are realized in the RR module:

- Monitoring of BCCH and PCH (readout of system information and paging messages)
- RACH administration: mobile stations send their requests for connections and replies to paging announcements to the BSS
- Requests for and assignments of data and signaling channels
- Periodic measurement of channel quality (quality monitoring)
- Transmitter power control and synchronization of the MS
- Handover (part of which is sometimes erroneously attributed to roaming functions and mobility management), always initiated by the network
- Synchronization of encryption and decryption on the data channel

The RR sublayer provides several services at the RR-SAP to the MM sublayer. These services are needed to set up and take down signaling connections and to transmit signaling messages.

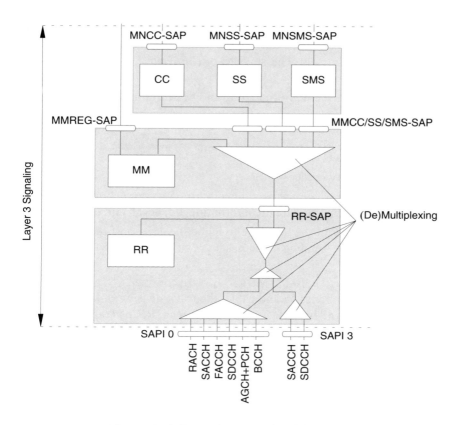

Figure 7.12: Layer 3 protocol architecture at the MS side

Mobility management (MM) encompasses all the tasks resulting from mobility. The MM activities are exclusively performed in cooperation between MS and MSC, and they include

- TMSI assignment
- Localization of the MS
- Location updating of the MS; parts of this are sometimes known as roaming functions
- Identification of the MS (IMSI, IMEI)
- Authentication of the MS
- IMSI attach and detach procedures (e.g. at insertion or removal of SIM)
- Ensuring confidentiality of subscriber identity

Registration services for higher layers are provided by Layer 3 at the MMREG-SAP (Figure 7.12). Registration involves the IMSI attach and detach procedures which are used by the mobile to report state changes such as power-up or power-down, or SIM card removal or insertion.

The MM sublayer offers its services at the MMCC-SAP, MMSS-SAP, and MMSMS-SAP to the CC, SS, and SMS entities. This is essentially a connection to the network side over which these units can communicate.

Connection management consists of three entities: *Call Control* (CC), *Supplementary Servi- ces* (SS), and *Short Message Service* (SMS). *Call control* handles all tasks related to setting up, maintaining and taking down calls. The services of *call control* are provided at the MNCC-SAP, and they encompass:

- Establishment of normal calls (MS-originating and MS-terminating)
- Establishment of emergency calls (only MS-originating)
- Termination of calls
- *Dual-Tone Multifrequency* (DTMF) signaling
- Call-related supplementary services
- Incall modification: the service may be changed during a connection (e.g. speech and transparent/nontransparent data are alternating; or speech and fax alternate)

The service primitives at this SAP of the interface to higher layers report reception of in- coming messages and effect the sending of messages, essentially ISDN user-network signaling according to Q.931.

RR messages are mainly exchanged between MS and BSS. In contrast, CM and MM func- tions are handled exclusively between MS and MSC; the exact division of labor between BTS, BSC, and MSC is summarized in Table 7.1. As can be seen, RR messages have to be transported over the Um and Abis interfaces, whereas CM and MM messages need addi- tional transport mechanisms across the A interface.

From a conceptual viewpoint, the A interface in GSM networks is the interface between the MSCs, the ISDN exchanges with mobile network specific extensions, and the BSC, the dedicated mobile network specific control units. Here too is the reference point, where the signaling system changes from GSM-specific to the general ISDN-compatible SS#7. Message transport in the SS#7 network is realized through the *Message Transfer Part* (MTP). In essence, MTP comprises the lower three layers of the OSI Reference Model, i.e. the MTP provides routing and transport of signaling messages.

A slightly modified (reduced) version of the MTP, called MTP ', has been defined for the protected transport of signaling messages across the A interface between BSC and MSC. At the ISDN side of the MSCs, the complete MTP is available. For signaling transactions between MSC and MS (CM, MM), it is necessary to establish and identify distinct logical connections. The *Signaling Connection Control Part* (SCCP) is used for this purpose to fa- cilitate implementation with a slightly reduced range of functions defined in SS#7.

For GSM-specific signaling between MSC and BSC, the *Base Station System Application Part* (BSSAP) has been defined. The BSSAP consists of the *Direct Transfer Application Part* (DTAP) and the *Base Station System Management Part* (BSSMAP). The DTAP is used to transport messages between MSC and MS. These are the *Call Control* (CC) and *Mobility Management* (MM) messages. At the A interface, they are transmitted with DTAP and then passed transparently through the BSS across the Abis interface to the MS without interpretation by the BTS.

The BSSMAP is the protocol definition part which is responsible for all of the administra- tion and control of the radio resources of the *Base Station Subsystem* (BSS). RR is one of the main functions of a BSS. Therefore, the RR entities terminate in the mobile station and the BTS or BSC respectively. Some functions of RR however, require involvement of the MSC (e.g. some handover situations, or release of connections or channels). Such actions should be initiated and controlled by the MSC (e.g. handover and channel assignment). This control is the responsibility of BSSMAP. RR messages are mapped and converted within the BSC into procedures and messages of BSSMAP and vice versa. BSSMAP offers the functions which are required at the A interface between BSS and MSC for RR of the BSS. Accordingly, RR messages initiate BSSMAP functions, and BSSMAP functions control RR protocol functions.

Table 7.1: Distribution of functions between BTS, BSC, and MSC
(according to GSM Rec. 08.02 and 08.52)

	BTS	BSC	MSC
Terrestrial Channel Management			
MSC-BSC-Channels			
Channel allocation			X
Blocking indication		X	
BSC-BTS-Channels			
Channel allocation		X	
Blocking indication	X		
Mobility Management			
Authentication			X
Location Updating			X
Call Control			X
Radio Channel Management			
Channel Coding/Decoding	X		
Transcoding/Rate Adaptation	X		
Interworking Function			X
Measurements			
Uplink measuring	X		X
Processing of reports from MS/TRX	X	X	X
Traffic measurements			X
Handover			
BSC internal, intracell		X	
BSC internal, intercell		X	
BSC external		X	
Recognition, decision, execution			X
HO access detection	X		
Paging			
Initiation		X	
Execution	X		

Table 7.1: Continued

	BTS	BSC	MSC
Radio Channel Management			
Channel Configuration Management		X	
Frequency Hopping			
Management		X	
Execution	X		
TCH Management			
Channel allocation		X	
Link supervision		X	
Channel release		X	X
Idle channel observation	X		
Power control determination	X	X	
SDCCH Management			
SDCCH allocation		X	
Link supervision		X	
Channel release		X	X
Power control determination	X	X	
BCCH/CCCH Management			
Message scheduling management		X	
Message scheduling execution	X		
Random access detection	X		
Immediate assign		X	
Timing Advance			
Calculation	X		
Signalling to MS at random access		X	
Signalling to MS at handover/during call	X		
Radio Resource Indication			
Report status of idle channels	X		
LAPDm Functions	X		
Encryption			
Management		X	
Execution	X		

A similar situation exists at the Abis interface. Most of the RR messages are passed transparently by the BTS between MS and BSC. Certain RR information, however, must be interpreted by the BTS, e.g. in situations like random access of the MS, the start of the ciphering process, or paging to localize an MS for connection setup. The *Base Transceiving*

Station Management (BTSM) contains functions for the treatment of these messages and other procedures for BTS management. Besides, a mapping occurs in the BTS from BTSM onto the RR messages relevant at the air interface (RR ', Figure 7.11).

The MSC is equipped with the *Mobile Application Part* (MAP), a mobile network specific extension of SS#7, for communication with the other components of the GSM network (the HLR and VLR registers, other MSCs) and other PLMNs. Among the MAP functions are all signaling functions among MSCs as well as between MSC and the registers (Figure 7.13). These functions include

- Updating of residence information in the VLR
- Cancellation of residence information in the VLR
- Storage of routing information in the HLR
- Updating and supplementing of user profiles in HLR and VLR
- Inquiry of routing information from the HLR
- Handover of connections between MSC

The exchange of MAP messages, e.g. with other MSCs, HLR, or VLR, occurs over the transport and transaction protocol of the SS#7. The SS#7 transaction protocol is the *Transaction Capabilities Application Part* (TCAP). A connectionless transport service is offered by the *Signaling Connection Control Part* (SCCP).

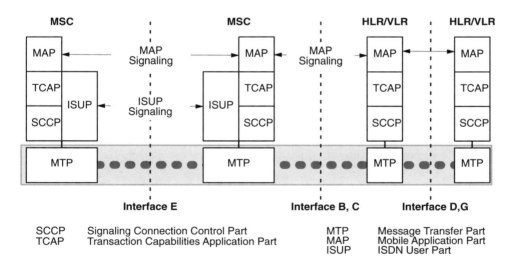

Figure 7.13: Protocol interfaces in the mobile network

The MAP functions require channels for signaling between different PLMNs which are provided by the international SS#7. Access to SS#7 occurs through the fixed ISDN.

Connection to the fixed network is typically done through leased lines; in the case of the German GSM network operators, it is through lines with a rate of 2 Mbit/s from Deutsche Telekom [9]. Often the majority of the MSC in a PLMN has such an access to the fixed network. On these lines, both user data and signaling data is transported. From the viewpoint of a fixed network, an MSC is integrated into the network like a normal ISDN exchange node. Outside of a PLMN, starting with the GMSC, calls for mobile stations are treated like calls for subscribers of the fixed network, i.e. the mobility of a subscriber with an MSISDN becomes "visible" only beyond the GMSC. For CC, the MSC has the same interface as an ISDN switching node. Connection-oriented signaling of GSM networks is mapped at the fixed network side (interface to ISDN) into the *ISDN User Part* (ISUP) used to connect ISDN channels through the network (Figure 7.14). The mobile-specific signaling of the MAP is routed over a gateway of the PLMN (GMSC) and the *International Switching Center* (ISC) of the national ISDN network into the international SS#7 network [9]. In this way, transport of signaling data between different GSM networks is also guaranteed without problems.

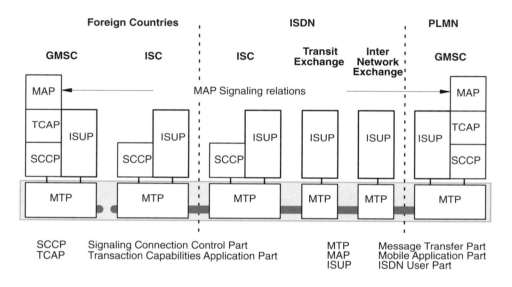

Figure 7.14: International signaling relations via ISDN [9]

7.3.2   Transport of User Data in the Signaling Plane

In the signaling plane (control plane) of the GSM architecture, one can also transport packet-oriented user data from or to mobile stations. This occurs for the point-to-point SMS (see Section 4.2). Short messages are always transmitted in store-and-forward mode through a *Short Message Service Center* (SMS-SC). The service center accepts these messages, which can be up to 160 characters long, and forwards them to the recipients (other mobile stations or fax, email, etc.). In principle, GSM defines a separate protocol architecture for the realization of this service.

Between mobile station and service center, short messages are transmitted using a connectionless transport protocol: *Short Message Transport Protocol* (SM-TP) which uses the services of the signaling protocols within the GSM network. Transport of these messages outside of the GSM network is not defined. For example, the SMS-SC could be directly connected to the gateway switching center (SMS-GMSC), or it could be connected to a *Short Message Service Interworking MSC* (SMS-IWMSC) through an X.25 connection (Figure 7.15). Within the GSM network between MSCs, a short message is transferred with the *Mobile Application Part* (MAP) and the lower layers of SS#7. Finally, between a mobile station and its local MSC, two protocol layers are responsible for the transfer of transport protocol units of SMS. First, there is the SMS entity in the CM sublayer of Layer 3 at the user-network interface (see Figure 7.12) which realizes the *Short Message Control Protocol* (SM-CP) and its connection-oriented service. Second, there is the relay layer, in which the *Short Message Relay Protocol* (SM-RP) is defined, which offers a connectionless service for transfer of SMS transport PDUs between MS and MSC. This, however, uses services at the service access point MMSMS-SAP (see Figure 7.12) and thus a connection of the MM sublayer.

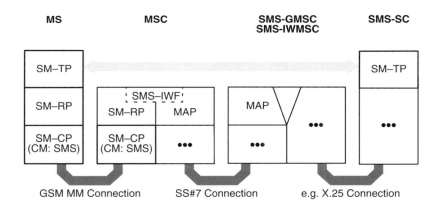

Figure 7.15: Protocol architecture for SMS transfer

In addition to the SM-CP, the relay protocol SM-RP was introduced above the CM sub-layer (Figure 7.12 and Figure 7.15) to realize an acknowledged transmission of short messages, but with minimal overhead for the radio channel. A short message sent by a mobile station is passed over the signaling network until it reaches the service center SMS-SC. If the service center determines the error-free reception of a message, an acknowledgment message is returned on the reverse path, which finally causes sending of an acknowledgment message from the SM-RP entity in the MSC to the mobile station. Until this acknowledgment message arrives, the connection in the MM sublayer can be taken down, and thus also the reserved radio channel. In this way, radio resources across the air interface are only occupied during the actual transmission of SM-RP messages. And each successful transmission of an SM-CP PDU across the MM connection, which includes the error-prone air interface, is immediately acknowledged, or else errors are immediately reported to the sending SM-CP entity. So if a message is damaged at the radio interface, this avoids it being transmitted to the service center.

7.4   Signaling across the User Interface (Um)

Signaling at the User–Network Interface in GSM is essentially concentrated in Layer 3. Layers 1 and 2 provide the mechanisms for the protected transmission of signaling messages across the air interface. Besides the local interface, they contain functionality and procedures for the interface to the BTS.

The signaling of Layer 3 at the User–Network Interface is very complex and comprises protocol entities in the mobile station and in all functional entities of the GSM network (BTS, BSC, and MSC).

7.4.1   Layer 1 of the MS-BTS Interface

Layer 1 of the OSI Reference Model (physical layer) contains all the functions necessary for the transmission of bit streams over the physical medium, in this case the radio channel. GSM Layer 1 defines a series of logical channels based on the channel access procedures with their physical channels. The higher layer protocols access these services at the Layer 1 service interface. The three interfaces of Layer 1 are schematically illustrated in Figure 7.16.

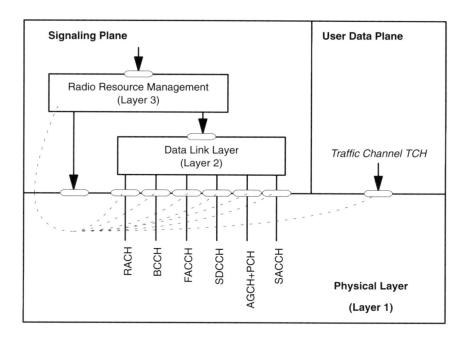

Figure 7.16: Layer 1 service interfaces

LAPDm protocol frames are transmitted across the service mechanisms of the data link layer interface, and the establishment of logical channels is reported to Layer 2. The communication across this interface is defined by abstract physical layer service primitives. A separate *Service Access Point* (SAP) is defined for each logical control channel (BCCH, PCH+AGCH, RACH, SDCCH, SACCH, FACCH).

Between Layer 1 and the RR sublayer of Layer 3 there is a direct interface. The abstract service primitives exchanged at this interface mostly concern channel assignment and Layer 1 system information, including measurement results of channel monitoring. At the third Layer 1 interface, the traffic channels for user (payload) data are provided.

The service access points (SAP) of Layer 1 as defined in GSM are not genuine service access points in the spirit of OSI. They differ from the PHY-SAPs of the OSI Reference Model insofar as these SAPs are controlled by Layer 3 RR sublayer (layer management, establishment and release of channels) rather than by control procedures in the link layer. Control of Layer 1 SAPs by RR comprises activation and deactivation, configuration, routing and disconnection of physical and logical channels. Furthermore, exchange of measurement and control information for channel monitoring occurs through service primitives.

7.4.1.1   Layer 1 Services

Layer 1 services of the GSM User–Network Interface are divided into three groups:

*   Access capabilities
*   Error detection
*   Encryption

Layer 1 provides a bit transport service for the logical channels. These are transmitted in multiplexed format over physical channels which consist of elements defined for the transmission on the radio channel (frequency, time slot, hopping sequence, etc.; see Section 7.1). Some physical channels are provided for common (shared) use (BCCH and CCCH), whereas others are assigned to dedicated connections with single mobile stations (dedicated physical channels). The combination of logical channels used on a physical channel can vary over time, e.g. TCH+SACCH/FACCH replaced by SDCCH+SACCH (see Table 5.4).

The GSM standard distinguishes explicitly between *access capabilities* for dedicated physical channels and for common physical channels BCCH/CCCHs. Dedicated physical channels are established and controlled by Layer 3 RR management. During the operation of a dedicated physical channel, Layer 1 continuously measures the signal quality of the used channel and the quality of the BCCH channels of the neighboring base stations. This measurement information is passed to Layer 3 in measurement service primitives MPH. In idle mode, Layer 1 selects the cell with the best signal quality in cooperation with the RR sublayer based on the quality of the BCCH/CCCH (cell selection).

GSM Layer 1 offers an error-protected bit transport service and therefore also error detection and correction mechanisms. To do this, error-correcting and error-detecting coding mechanisms are provided (see Section 6.2). Frames recognized as faulty are not passed up to Layer 2. Furthermore, security-relevant functions like encryption of user data is implemented in Layer 1 (see Section 6.3).

7.4.1.2   Layer 1: Procedures and Peer-to-Peer Signaling

GSM defines and distinguishes between two operational modes of a mobile station: *idle mode* and *dedicated mode* (Figure 7.17). In idle mode, the mobile station is either powered off (state NULL) or it searches for or measures the BCCH with the best signal quality (state

SEARCHING BCH), or is synchronized to a specific base station's BCCH and ready to perform a random access procedure on the RACH for requesting a dedicated channel in state BCH (see Section 5.5.4).

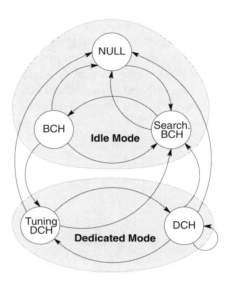

Figure 7.17: State diagram of a mobile station's physical layer

In state TUNING DCH of the *dedicated mode*, the mobile station occupies a physical channel and tries to synchronize with it, which will eventually result in transition to state DCH. In this state, the MS is finally ready to establish logical channels and switch them through. The state transitions of Layer 1 are controlled by MPH service primitives of the RR interface, i.e. directly from the Layer 3 RR sublayer of the signaling protocol stack.

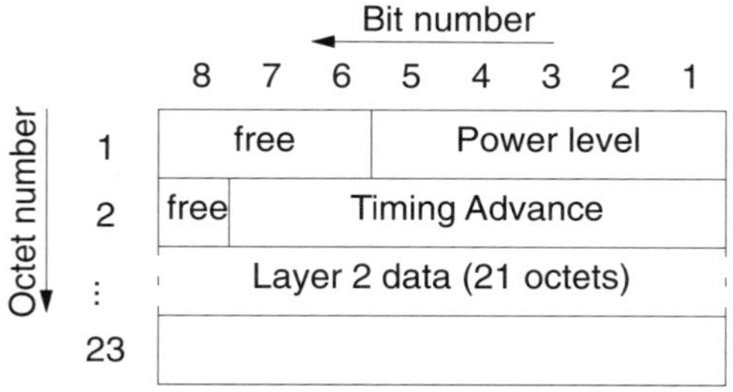

Figure 7.18: Format of an SACCH block

Layer 1 defines its own frame structure for the transport of signaling messages, which occur as LAPDm frames at the SAP of the respective logical channel. Figure 7.18 shows the format of an SACCH block as an example, which essentially contains 21 octets of LAPDm data.

Furthermore, the SACCH frame contains a kind of protocol header which carries the current power level and the value of the timing advance. This header is omitted in the other logical channels (FACCH, SDCCH, CCCH, BCCH) which contain only LAPDm PDUs.

7.4.2 Layer 2 Signaling

The LAPDm protocol is the data link protocol for signaling channels at the air interface. It is similar to HDLC. It provides two operational modes:

- Unacknowledged operation
- Acknowledged operation

In the *unacknowledged operation* mode, data is transmitted in UI frames (unnumbered information) without acknowledgment; there is no flow control or L2 error correction. This operational mode is allowed for all signaling channels, except for the RACH which is accessed in multiple access mode without reservation or protection.

The *acknowledged operation* mode provides protected data service. Data is transmitted in I frames (information) with positive acknowledgment. Error protection through retransmission (ARQ) and flow control are specified and activated in this mode. This mode is only used on DCCH channels.

In LAPDm, the *Connection End Points* (CEPs) of L2 connections are labeled with *Data Link Connection Identifiers* (DLCIs), which consist of two elements:

- The Layer 2 *Service Access Point Identifier* (SAPI) is transmitted in the header of the L2 protocol frame.
- The physical channel identifier on which the L2 connection is or will be established, is the real Layer 2 *Connection End Point Identifier* (CEPI). The CEPI is locally administered and not communicated to the L2 peer entity. (The terminology of the GSM standard is somewhat inconsistent in this case − what is really meant is the respective logical channel. The physical channels from the viewpoint of LAPDm are the logical channels of GSM, rather than the physical channels defined by frequency/time slot/hopping sequence.)

When a Layer 3 message is transmitted, the sending entity chooses the appropriate SAP and CEP. When the service data unit SDU is handed over at the SAP, the chosen CEP is given to the L2 entity. Conversely, when receiving an L2 frame, the appropriate L2-CEPI can be determined from the physical / logical channel identity and the SAPI in the frame header.

Specific SAPI values are reserved for the certain functions:

- SAPI=0 for signaling (CM, MM, RR)
- SAPI=3 for SMS

In the control plane, these two SAPI values serve to separate signaling messages from packet-oriented user data (short messages). Further functions needing a new SAPI value can be defined in future versions of the GSM standard.

Table 7.2: Logical channels, operational modes and Layer 2 SAPIs

Logical Channel	SAPI=0	SAPI=3
BCCH	Unacknowledged	–
CCCH	Unacknowledged	–
SDCCH	Unacknowledged and Acknowledged	Unacknowledged and Acknowledged
SACCH assoc. with SDCCH	Unacknowledged	–
SACCH assoc. with TCH	Unacknowledged	Unacknowledged and Acknowledged
FACCH	Unacknowledged and Acknowledged	–

An LAPDm entity is established for each of the pertinent physical / logical channels. For some of the channel / SAPI combinations only a subset of the LAPDm protocol is needed (e.g. unacknowledged operation), and some channel / SAPI combinations are not supported (Table 7.2). These LAPDm entities perform the *Data Link* procedure, i.e. the functions of the L2 peer-to-peer communication as well as the service primitives between adjacent layers. Segmentation and reassembly of Layer 3 messages is also included.

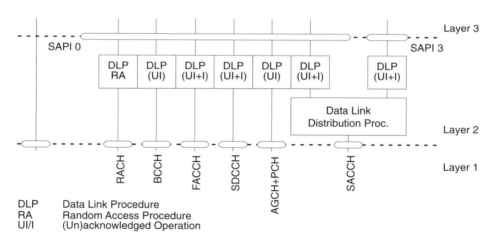

Figure 7.19: Sample configuration of the MS data link layer

Further Layer 2 procedures are the *Distribution Procedure* and the *Random Access* (RA) procedure. The distribution procedure is needed if multiple SAPs are associated with one physical / logical channel. It performs the distribution of the L2 frames received on one channel to the respective data link procedure, or the priority-controlled multiplexing of L2 frames from multiple SAPs onto one channel. The random access procedure is used on the random access channel (RACH); it deals with the random controlled retransmission of random access bursts, but it does not perform any error protection on the unidirectional RACH.

For certain aspects of RR, the protocol logic of Layer 3 has to have direct access to the services of Layer 1. Especially, this is needed for functions of *Radio Subsystem Link Control,* i.e. for channel measurement, transmitter power control, and timing advance.

A possible link layer configuration of an MS is shown in Figure 7.19. The base station has a similar configuration with one PCH+AGCH, SDCCH and SACCH/FACCH for each active mobile station.

Figure 7.20 shows the different types of protocol data frames used for communication between L2 peer entities in MS and BS. Frame formats A and B are used on the SACCH, FACCH and SDCCH channels, depending upon whether the frame has an information field (Type B) or not (Type A). For unacknowledged operation (BCCH, PCH, AGCH), format types Abis and Bbis are used on channels with SAPI=0. The Abis format is used when there is no information to be transmitted on the respective logical channel.

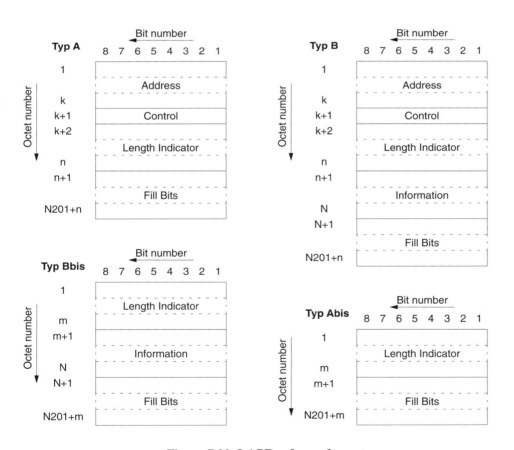

Figure 7.20: LAPDm frame formats

In contrast to HDLC, LAPDm frames have no flag to designate beginning and end of a frame, rather the delineation of frames is done as in RLP at the link level (see Section 7.2.3) through the fixed-length block structure of Layer 1. The maximum number of octets N201 per information field depends on the type of logical channel (Table 7.3). The end of the information field is given by a *Length Indicator*, a value of less than N201 indicates that the frame has to be supplemented with fill bits to the full length. In case of an SACCH channel, for example, this yields a fixed-length LAPDm packet of 21 octets. Combined with the fields for transmitter power control and timing advance, an SACCH block of Layer 1 is thus 23 octets long.

The address field may have a variable length, however; for use on control channels it consists of exactly one octet. Besides other fields, this octet contains an SAPI (3 bits) and the *Command/Response* (C/R) flag known from HDLC. In LAPDm, the coding of the control field with sending and receiving sequence numbers and the state diagram describing the protocol procedures are almost identical to HDLC [15]. Some additional parameters are required at the service interface to Layer 3; for example, a parameter CEP designating the desired logical channel.

Furthermore, the LAPDm protocol has some simplifications or peculiarities with regard to HDLC:

- The sending window size is restricted to $k=1$.
- The protocol entities should be implemented in such a way that the state RECEIVER BUSY is never reached. Thus RNR packets can be safely ignored. The HDLC polling procedure for state inquiry of the partner station need not be implemented in LAPDm.
- Connections to SAPI=0 are always initiated by the mobile station.

In addition, the repetition timer T200 and the maximum number of allowed repetitions N200 have been adapted to the special needs of the mobile channel. In particular, they have their own value determined by the type of logical channel.

Table 7.3: Logical channels and the maximum length of the LAPDm information field

Logical channel	N201
SACCH	18 octets
SDCCH, FACCH	20 octets
BCCH, AGCH, PCH	22 octets

7.4.3   Radio Resource Management

The procedures for *Radio Resource Management* (RR) are the basic signaling and control procedures at the air interface. They handle the assignment, allocation and administration of radio resources, the acquisition of system information from broadcast channels (BCCH) and the selection of the cell with the best signal reception (see cell selection in Section 5.5.4.1). Accordingly, the RR procedures and pertinent messages (Table 7.4) are defined for idle mode as well as for setting up, maintaining, and taking down of RR connections. Figure 7.21 shows the format of RR messages, which is uniform for all three Layer 3 signaling sublayers (CM, MM, RR).

Each Layer 3 message contains a protocol discriminator in the first octet, which allows association of messages with the respective sublayer or service access point (Figure 7.12). The uppermost four bits of the first octet also contain a *Transaction ID*, which enables an

MS to perform several signaling transactions in parallel. The *Message Type* (MT) is registered in the lower seven bits of the second octet (see also Table 7.4, Table 7.5 and Table 7.6). Otherwise, Layer 3 messages consist of *Information Elements* (IEs) of fixed or variable length; a *Length Indicator* (LI) is added for variable-length messages.

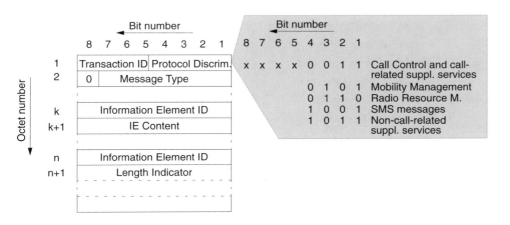

Figure 7.21: Format of a Um signaling message (Layer 3)

In idle mode, the MS is reading continuously the BCCH information and conducts periodic measurements of the signaling strength of the BCCH carriers in order to be able to select the current cell (see Section 5.5.4). In this state, there is no exchange of signaling messages with the network. The data required for RR and other signaling procedures is collected and stored: the list of neighboring BCCH carriers, thresholds for RR algorithms, CCCH configurations, information about the use of RACH and PCH, etc. This information is broadcast by the BSS on the BCCH (SYSTEM INFORMATION, Type 1 through 4) and therefore is available to all mobile stations currently in the cell. Also important is the periodic monitoring of the paging channel (PCH) so that paging calls are not lost. For this purpose, the BSS is sending on all paging channels of a cell continuously valid Layer 3 messages (PAGING REQUEST) which the MS can decode and recognize if its address is paged.

Each exchange of signaling messages with the network (BSS, MSC) requires an RR connection and the establishment of an LAPDm connection between MS and BTS. Setting up the RR connection can be initiated by the network or the MS (Figure 7.22). In either case, the MS sends a channel request (CHANNEL REQUEST) on the RACH in order to get a channel assigned on the AGCH (immediate assignment procedure). There is also a procedure to deny a channel request (immediate assignment reject).

If the network does not immediately answer to the channel request, the request is repeated using the Aloha method with a random number controlled timer (Figure 7.22). In the case of a network-initiated connection, this procedure is preceded by a paging call (PAGING REQUEST) to be answered by the mobile station (PAGING RESPONSE). After an RR connection has been successfully completed, the higher protocol layers (CM, MM) can receive and transmit signaling messages at SAPI 0.

Table 7.4: RR messages

Category	Message	Logical channel	Direction	MT-Code
Channel	Additional Assignment	DCCH	N -> MS	00111011
Establishment	Immediate Assignment	CCCH	N -> MS	00111111
	Immediate Assignment Extended	CCCH	N -> MS	00111001
	Immediate Assignment Rejected	CCCH	N -> MS	00111010
Ciphering	Ciphering Mode Command	DCCH	N -> MS	00110101
	Ciphering Mode Complete	DCCH	MS -> N	00110010
Handover	Assignment Command	DCCH	N -> MS	00101110
	Assignment Complete	DCCH	MS -> N	00101001
	Assignment Failure	DCCH	MS -> N	00101111
	Handover Access	DCCH	MS -> N	–
	Handover Command	DCCH	N -> MS	00101011
	Handover Complete	DCCH	MS -> N	00101100
	Handover Failure	DCCH	MS -> N	00101000
	Physical Information	DCCH	N -> MS	00101101
Channel	Channel Release	DCCH	N -> MS	00001101
Release	Partial Release	DCCH	N -> MS	00001010
	Partial Release Complete	DCCH	MS -> N	00001111
Paging	Paging Request, Type 1/2/3	PCH	N -> MS	00100xxx
	Paging Response	DCCH	MS -> N	00100111
System	System Information Type 1/2/3/4	BCCH	N -> MS	00011xxx
Information	System Information Type 5/6	SACCH	N -> MS	00011xxx
Miscellaneous	Channel Mode Modify	DCCH	N -> MS	00010000
	Channel Mode Modify Acknowledge	DCCH	MS -> N	00010111
	Channel Request	RACH	MS -> N	–
	Classmark Change	DCCH	MS -> N	00010110
	Frequency Redefinition	DCCH	N -> MS	00010100
	Measurement Report	SACCH	MS -> N	00010101
	Synchronization Channel Information	SCH	N -> MS	–
	RR-Status	DCCH	MS <-> N	00010010

In contrast to the setup of connections, the release is always initiated by the network (CHANNEL RELEASE). Reasons for the release of the channel could be end of the signaling transaction, too many errors, removal of the channel in favor of a higher priority call (e.g. emergency call), or end of a call. After receiving the channel release command, the mobile station assumes the idle state following a brief waiting period (Figure 7.22).

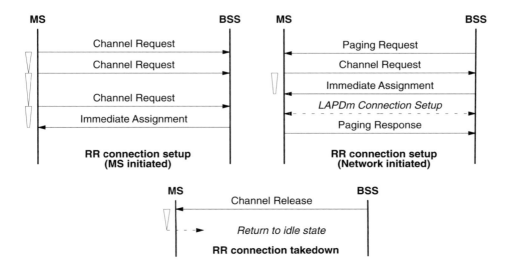

Figure 7.22: RR connection setup and takedown

Once an RR connection has been set up, the mobile station has either an SDCCH or a TCH with associated SACCH/FACCH available for exclusive bidirectional use. On the SACCH, data must be sent continuously (see also Section 5.5.3), i.e. the MS keeps sending current channel measurements (MEASUREMENT REPORT, see Section 5.5.1.2) if no other signaling messages need to be sent. In the other direction, the BSS keeps sending system information (SYSTEM INFORMATION, alternating between Type 5 and Type 6). The information element with the coded measurement results contains the following among other data: RXLEV and RXQUAL of the serving cell as well as RXLEV and carrier frequency of up to six neighboring cells as well as their BSICs (Figure 7.23). The system information sent by the BSS on the SACCH contains first information about the neighbor cells and their BCCH (Type 5), and second, information about the current cell (Type 6) such as *Cell Identity* (CI) and the current *Location Area Identifier* (LAI).

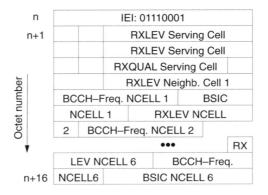

Figure 7.23: Measurement result (information element)

For established RR connections, a channel change within the cell can be performed (dedicated channel assignment, Figure 7.24) to change the configuration of the physical channel in use. Such a channel change can be requested by higher protocol layers, or it can be requested by the RR sublayer; however, it is always initiated by the network. When the mobile station receives an ASSIGNMENT COMMAND, the transmission of all signaling messages is suspended, the LAPDm connection is taken down, the traffic channel, if existent, is switched off, and the old channel is deactivated. After activation of the new physical channel and a successful establishment of a new LAPDm connection (Layer 2), the held-back signaling messages can be transmitted.

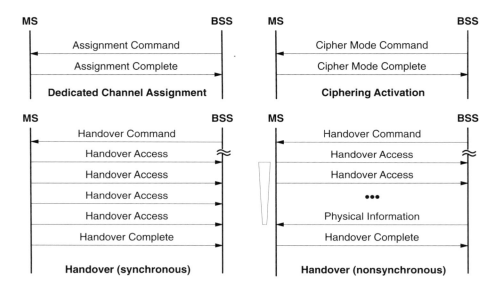

Figure 7.24: Channel change, encryption, and handover

A second signaling procedure to change the physical channel configuration of an established RR connection is the handover procedure, which is also initiated only from the network side and, for example, becomes necessary if the current cell is left. In contrast to AS-SIGNMENT COMMAND, a HANDOVER COMMAND contains not only the new channel configuration but also information about the new cell (e.g. BSIC and BCCH frequency), the procedure variant to establish a physical channel (asynchronous or synchronous handover, Figure 7.24), and a handover reference number.

Having received a HANDOVER COMMAND on the FACCH, the mobile station terminates the LAPDm connection on the old channel, interrupts the connection, deactivates the old physical channel, and finally switches over to the channel newly assigned in the HANDOVER COMMAND. On the main DCCH (in this case FACCH), the mobile station sends the unencrypted message HANDOVER ACCESS in an access burst (Figure 5.6, coding as on the RACCH, see Section 6.2) to the base station. Even though this is a message on the FACCH, an access burst is used because the mobile station at this time does not yet know the complete synchronization information. The eight data bits of the access burst contain the handover reference of the handover command. The way in which these access bursts are transmitted depends on whether both cells have synchronized their TDMA transmission or not. In the case of existing synchronization, the access burst (HANDOVER ACCESS) is sent in exactly four subsequent time slots of the main DCCH (FACCH). Thereafter, the mobile station activates the new physical channel in both directions, establishes an LAPDm connection, activates encryption, and finally sends a message HANDOVER COMPLETE to the BSS. In the nonsynchronous case, the mobile station repeats the access burst until either a timer expires (handover failed) or until the base station answers with an RR message PHYSICAL INFORMATION which contains the currently needed timing advance and this way enables the establishment of the new RR connection.

Another important RR procedure is the activation of ciphering. This is done by the BSS with the CIPHER MODE COMMAND, which also indicates that the BTS has activated its deciphering function. Having received the CIPHER MODE COMMAND, the mobile station activates ciphering as well as deciphering and sends the answer CIPHER MODE COMPLETE already in enciphered form. If the BTS is able to correctly decipher this message, the ciphering mode has been successfully established.

In addition, there are a number of less significant signaling procedures defined, such as *Frequency Redefinition, Additional Assignment, Partial Release,* or *Classmark Change.* The first one concerns the change of the *Mobile Allocation* (MA); see Section 5.2.3. The next two deal with a change of the physical channel configuration. With the last message, CLASSMARK CHANGE, the mobile station reports that it now belongs to a new power class (see Table 5.8), which can be achieved by installing a commercially available power booster kit, for example.

7.4.4 Mobility Management

The main task of *Mobility Management* (MM) is to support the mobility of the mobile station; for example, by reporting the current location to the network or verifying the subscriber identity. Another task of the MM sublayer is to offer MM connections and associated services to the CM sublayer above. The message format for MM messages is the uniform Layer 3 signaling message format (Figure 7.21). MM has its own protocol discriminator, and the MM messages are marked with a type code (MT, Table 7.5).

All MM procedures presume an established RR connection, i.e. a dedicated logical channel must be assigned with an established LAPDm connection in place, before MM transactions can be performed. These transactions occur between MS and MSC, i.e. messages are passed through the BSS transparently without interpretation and forwarded to the MSC with the DTAP transport mechanism. The MM procedures are divided into three categories: *Common, Specific,* and *MM Connection Management.* Whereas *Common* procedures can always be initiated and executed as soon as an RR connection exists, *Specific* procedures exclude one another, i.e. they cannot be processed simultaneously or during an MM connection. Conversely, an MM connection can only be set up if no *Specific* procedure is running.

Table 7.5: MM messages

Category	Message	Direction	MT
Registration	IMSI Detach Indication	MS -> N	0x000001
	Location Updating Accept	N -> MS	0x000010
	Location Updating Reject	N -> MS	0x000100
	Location Updating Request	MS -> N	0x001000
Security	Authentication Reject	N -> MS	0x010001
	Authentication Request	N -> MS	0x010010
	Authentication Response	MS -> N	0x010100
	Identity Request	N -> MS	0x001000
	Identity Response	MS -> N	0x001001
	TMSI Reallocation Command	N -> MS	0x001010
	TMSI Reallocation Complete	MS -> N	0x001011
Connection Management	CM Service Accept	MS <-> N	0x100001
	CM Service Reject	N -> MS	0x100010
	CM Service Request	MS -> N	0x100100
	CM Reestablishment Request	MS -> N	0x101000
Miscellaneous	MM-Status	MS <-> N	0x110001

The *Common* MM procedures are summarized in Figure 7.25. Besides the *IMSI Detach* procedure, they are all initiated from the network side. An important role for the protection of subscriber identity is held by the *TMSI Reallocation* procedure. If the confidentiality of a subscriber's identity IMSI is to be protected (an optional network service), the signaling procedures across the air interface use the TMSI instead of the IMSI. This TMSI has only local significance within a *Location Area* and must be used together with the LAI for the unique identification of a subscriber.

For further protection, the TMSI can also be repeatedly reallocated (TMSI reallocation) which must be done at the latest when the location area changes. Otherwise this TMSI change is left as an option to the network operator, but it can be performed any time after an encrypted RR connection to the mobile station has been set up. The TMSI reallocation is either executed explicitly as a standalone procedure, or implicitly from other procedures using the TMSI, e.g. the *Location Update*. In case of explicit TMSI reallocation, the network sends a TMSI REALLOCATION COMMAND with the new TMSI and the current LAI on an encrypted RR connection to the mobile station (Figure 7.25).

The MS stores the TMSI and LAI in nonvolatile SIM storage and acknowledges it with the message TMSI REALLOCATION COMPLETE. If this message reaches the MSC before the timer expires, the timer is canceled, and the TMSI is valid. However, if the timer expires before the acknowledgment arrives, the procedure is repeated. If it fails a second time, the old as well as the new TMSI are barred for a certain time interval, and the IMSI is used for paging the mobile station. If the mobile station answers a paging call, TMSI reallocation is started again. Furthermore, the TMSI is assumed valid in spite of failed reallocation if it is used by the MS in subsequent transactions.

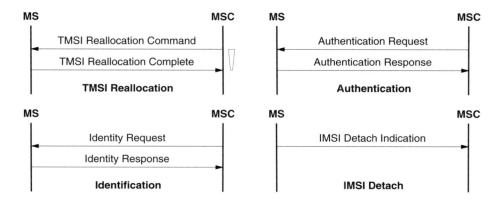

Figure 7.25: MM signaling procedures of category *Common*

Two more *Common* procedures are used for the identification of a mobile station or a subscriber (*Identification* procedure) and for the verification of the respective identity (*Authentication* procedure). For the identification of a mobile station, there is the equipment identity IMEI as well as the subscriber identity IMSI which is assigned to the MS through the SIM card. The network may request these two identity parameters at any time from the mobile station with an IDENTITY REQUEST . Therefore the mobile station must be able at any time to supply these identity parameters to the network with an IDENTITY RESPONSE message.

Authentication also assigns a new key for encryption of user payload data. This procedure is started from the network with an AUTHENTICATION REQUEST message. A mobile station must be able to process this request at any time during an RR connection. The MS calculates the new key Kc for the encryption of user data from the information obtained during authentication which is locally stored, and it also calculates authentication information to prove its identity without doubt. This authentication data is transmitted with an AUTHENTICATION RESPONSE message to the MSC which evaluates them. If the answer is not valid and the authentication has therefore failed, further processing depends on whether the IMSI or TMSI was used. In the case of TMSI, the network can start the identification procedure. If the implied IMSI is not identical to the one associated with the TMSI by the network, the authentication is restarted with new correct parameters. If the two IMSIs agree, or the IMSI was used a priori by the MS, the authentication has failed, which is indicated to the MS with an AUTHENTICATION REJECT message. This forces the MS to cancel all the assigned identity and security parameters (TMSI, LAI, Kc) and to enter idle mode, so that only simple cell selection and emergency calls are enabled.

If the mobile station is powered off or the SIM has been removed, the MS is not reachable because the MS does not monitor the paging channel, and calls cannot be delivered. In order to relieve the paging load on the BSS caused by unnecessary paging calls, a network operator can optionally request an explicit deregistration message from the mobile station, which is not normally required. This option is signaled by setting a flag on the BCCH (SYSTEM INFORMATION Type 3) and on the SACCH (SYSTEM INFORMATION Type 8). If the flag is set, the MS sends an IMSI DETACH INDICATION message when it powers off or when the SIM is removed, which allows the network to mark the MS as inactive. The IMSI detach procedure is the only *Common* procedure that cannot be started at an arbitrary time even during a *Specific* procedure. Its start has to be delayed until any *Specific* procedure has ended.

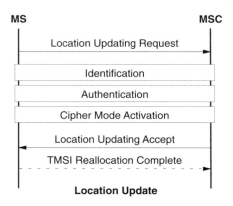

Figure 7.26: MM signaling procedures of category *Specific*

In GSM systems, updating of current location information is the sole responsibility of the mobile station. Using the information broadcast on BCCH channels, it has to recognize any change in the current location area and report it to the network, so that the databases HLR and VLR can be kept up to date. The generic structure of a location update is shown in Figure 7.26: The mobile station requests to update its current location information in the network with a LOCATION UPDATING REQUEST. If this can be done successfully, the network acknowledges this with a message LOCATION UPDATING ACCEPT. In the course of a location update, the network can ask for the mobile station's identity and check it out (identification and authentication). If the service "confidential subscriber identity" has been activated, a new TMSI assignment is a permanent component of the location update. In this case, enciphering of user data on the RR connection is activated, and the new TMSI is transmitted together with the message LOCATION UPDATING ACCEPT and is acknowledged with the message REALLOCATION COMPLETE. Periodic updating of location information can be used to indicate the presence of the mobile station in the network. For this purpose, the mobile station keeps a timer which periodically triggers a *Location Update* procedure. If this option is in use, the timer interval to be used is broadcast on the BCCH (SYSTEM INFORMATION Type 3). The procedure *IMSI Attach* is the converse of the procedure *IMSI Detach* (see Figure 7.25) and is executed as a special variant of the location update if the network requires this. However, the mobile station executes an *IMSI Detach* only if the LAI broadcast on the BCCH agrees with the LAI stored in the MS. If the stored LAI and received LAI differ, a normal *Location Update* procedure is executed.

Finally, there is a third category of MM procedures which are needed for the establishment and the operation of MM connections (Figure 7.27). An MM connection is established on request from the CM sublayer above and serves for the exchange of messages between CM entities, where each CM entity has its own MM connection (Figure 7.12). The procedures for the setup of MM connections are different depending on whether initiation occurs from the network or the mobile station.

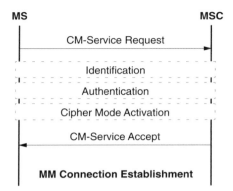

Figure 7.27: MM signaling procedures of category *MM Connection Management*

Setting up an MM connection from the side of the mobile station presumes the existence of an RR connection, but a single RR connection can be used by multiple MM connections. The MM connection can only be established if the mobile station has executed a successful location update in the current location area. An exception is an emergency call, which is possible at any time. If there is a request from the CM sublayer for an MM connection, it may be delayed or rejected if there are *Specific* procedures active, depending on implementation. If the MM connection can be established, the mobile station sends the message CM-SERVICE REQUEST to the network. This message contains information about the mobile station (IMSI or TMSI) as well as information about the requested service (outgoing voice call, SMS transfer, activation or registration of a supplementary service, etc.). Depending on these parameters, the network can execute any *Common* MM procedure (except *IMSI Detach*) or activate enciphering of user data. If the mobile station receives the message CM-SERVICE ACCEPT or the local message from the RR sublayer that enciphering was activated, it treats this as an acceptance of the service request, and the requesting CM entity is informed about the successful setup of an MM connection. Otherwise, if the service request has been rejected by the network, the MS receives a message CM-SERVICE REJECT, and the MM connection cannot be established.

The network-initiated setup of an MM connection does not require an exchange of CM service messages. After successful paging, an RR connection is established, and the sublayer on the network side executes one of the MM procedures if necessary (mostly *Location Update*) and requests from the RR sublayer the activation of user data encryption. If these transactions are successful, the service requesting CM entity is informed, and the MM connection is established.

7.4.5 Connection Management

Call Control (CC) is one of the entities of *Connection Management* (CM); the CM sublayer is shown in Figure 7.12. It comprises procedures to establish, control, and terminate calls. Several parallel CC entities are provided, such that several parallel calls on different MM connections can be processed. For CC, finite state models are defined both on the mobile side as well as on the network side. The two entities at the MS and MSC sites each instantiate a protocol automaton, and these communicate with each other using the messages in Table 7.6 and the uniform Layer 3 signaling message format (Figure 7.21).

Table 7.6: CC messages for circuit-switched connections

Category	Message	Direction	MT
Call Establishment	Alerting	N -> MS	0x000001
	Call Confirmed	MS -> N	0x001000
	Call Proceeding	N -> MS	0x000010
	Connect	N <-> MS	0x000111
	Connect Acknowledge	N <-> MS	0x001111
	Emergency Setup	MS -> N	0x001110
	Progress	N -> MS	0x000011
	Setup	N <-> MS	0x000101
Call Information Phase	Modify	N <-> MS	0x010111
	Modify Complete	N <-> MS	0x011111
	Modify Reject	N <-> MS	0x010011
	User Information	N <-> MS	0x010000
Call Clearing	Disconnect	N <-> MS	0x100101
	Release	N <-> MS	0x101101
	Release Complete	N <-> MS	0x101010
Miscellaneous	Congestion Control	N <-> MS	0x111001
	Notify	N <-> MS	0x111110
	Start DTMF	MS -> N	0x110101
	Start DTMF Acknowledge	N -> MS	0x110010
	Start DTMF Reject	N -> MS	0x110111
	Status	N <-> MS	0x111101
	Status Enquiry	N <-> MS	0x110100
	Stop DTMF	MS -> N	0x110001
	Stop DTMF Acknowledge	N -> MS	0x110010

Parts of CC in the mobile station are presented schematically in Figure 7.28 and Figure 7.29. They show the mobile-originating and mobile-terminating setup of a call and the mobile / network initiated call takedown. If there is a desire to call from the mobile station (mobile-originating call), the CC entity first requests an MM connection from the local MM entity, also indicating whether it is a normal or emergency call (MMCC Establishment Request, Figure 7.28). The call to be established on this MM connection requires a special service quality of the MM sublayer.

For a simple call, the mobile station must be registered with the network, whereas this is only optionally required with an emergency call, i.e. an emergency call can also be established on an unenciphered RR connection from a mobile station that is not registered.

After successful establishment of this MM connection and activation of the user data encryption, the service-requesting CC entity is informed (further interactions with the MM entity are not shown in Figure 7.28 for call establishment). The mobile station signals on this connection the desire to connect to the CC entity in the MSC (SETUP). An emergency call is initiated with the message EMERGENCY SETUP; the remaining call setup is identical to the one used for single calls.

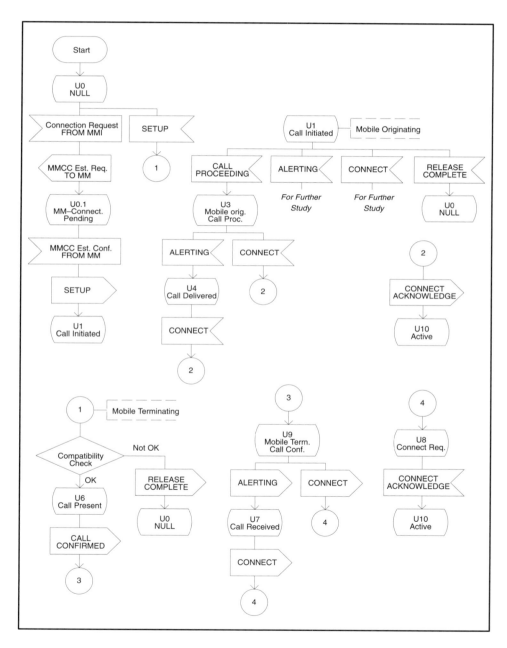

Figure 7.28: Call setup (mobile station): mobile-originating and mobile-terminating

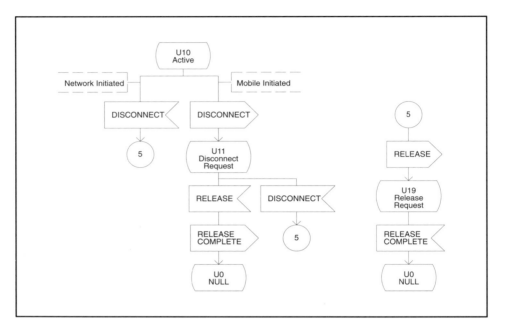

Figure 7.29: Call termination at the mobile station:
mobile-initiated and network-initiated

The MSC can respond to this connection request in several ways: it can indicate with a message CALL PROCEEDING that the call request has been accepted and that all the necessary information for the setup of the call is available. Otherwise the call request is declined with RELEASE COMPLETE. As soon as the called party is being signaled, the MS receives an ALERTING message; once the called party accepts the call, a CONNECT message is returned which is acknowledged with a CONNECT ACKNOWLEDGE message, thus switching through the call and the associated user data connection. If the call request need not be signaled to the called party and the call can be accepted directly, the ALERT message is omitted. Essentially CC signaling in GSM corresponds to the call setup according to Q.931 in ISDN. In addition, CC in GSM has a number of peculiarities, especially to account for the limited resources and properties of the radio channel. In particular, the call request of the MS can be entered into a queue (call queuing), if there is no immediately free traffic channel (TCH) for the establishment of the call. The maximum waiting time a call may have to wait for assignment of a traffic channel can be adjusted according to operator needs. Furthermore, the point at which the traffic channel is actually assigned can be chosen. For example, the traffic channel can be assigned immediately after acknowledging the call request (CALL PROCEEDING); this is *Early Assignment*. On the other hand, the call can be first processed completely and the assignment occurs only after the targeted subscriber is being called; this is *Late Assignment* or *Off-Air Call Setup* (OACSU). The variant OACSU avoids unnecessary allocation of a traffic channel if the called subscriber is not available. The blocking probability for call arrivals at the air interface can be reduced this way. On the other hand, there is the probability that after a successful call request signaling procedure, no traffic channel can be allocated for the calling party before the called party accepts the call, and thus the call cannot be completely switched through and must be broken off.

If a call arrives at the mobile station (mobile-terminating call), an RR connection with the mobile station is established within the MM connection setup (inclusive of paging). Once the MM connection is successfully completed and the encryption is activated, the call request is signaled to the mobile station with a SETUP message. This message includes information about the requested service, and the mobile station examines first whether it can satisfy the requested service profile (compatibility check). If affirmative, it accepts the call request and signals this to the local subscriber (local generation of call signal). This is finally communicated to the MSC with a CALL CONFIRMED message and an ALERTING message. If the mobile subscriber eventually accepts the call, the call is switched through completely with handshake messages CONNECT and CONNECT ACKNOWLEDGE. If because of the selected service there is no necessity for call request signaling to the called subscriber and the call can be switched through immediately (e.g. with fax call), the mobile station signals the call acceptance (CONNECT) immediately after the message CALL CONFIRMED. Call queuing and OACSU can also be used for mobile-terminating calls. The OACSU variant for mobile-terminating calls allocates a traffic channel only after the call has been accepted by the mobile subscriber with a CONNECT message.

The release of a connection is started with a DISCONNECT message either from the mobile or the network side (mobile-/network-initiated) and is completed with handshake messages RELEASE and RELEASE COMPLETE. If there is a collision of DISCONNECT messages, i.e. if both CC entities send a DISCONNECT simultaneously, they also answer it with a RELEASE so that a secure termination is ensured.

During an established call, two more CC procedures can be employed: *Dual-Tone Multifrequency* (DTMF) signaling and *Incall Modification*. DTMF signaling is an inband signaling procedure, which allows terminals (here mobile stations) to communicate with the respective other side, e.g. answering machines, or configuring special network services, e.g. voice mailboxes in the network. In GSM, DTMF can only be used during a voice connection. With a message START DTMF on the FACCH, the network is told that a key has been pressed, and the release of the key is signaled with a STOP DTMF message (Figure 7.30). Each of these messages is acknowledged by the network (MSC). A minimum interval must be maintained between the START/STOP messages (T_{press_min}, $T_{release\_min}$). While a key is depressed at the mobile station, the MSC generates a DTMF tone corresponding to the key code signaled with the START DTMF command. The DTMF tones must be generated within the MSC, since the speech coding in the GSM codec does not permit the pure transmission of DTMF tones in the voice band, and thus DTMF tones generated by the MS would arrive at the other side in distorted form.

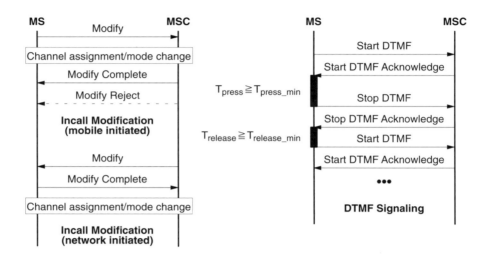

Figure 7.30: DTMF signaling and service change

Using the *Incall Modification* procedure, a service change can be performed, e.g. when speech and fax data is sent in sequence and are alternating during a call (see Chapter 4). A service change is started either from the mobile station or from the network by sending a MODIFY request. This request contains the service and kind of change (returning or non-returning). After sending the MODIFY request, the transmission of user data is halted. If the service change can be performed, this is signaled with a MODIFY COMPLETE message; otherwise the request is denied with a MODIFY REJECT. The service change may necessitate a change in the current physical channel configuration or operating mode. For this purpose, the MSC will use the respective channel assignment (ASSIGNMENT COMMAND, see Figure 7.24).

7.4.6 Structured Signaling Procedures

The preceding sections have presented the basic signaling procedures of the three sublayers RR, MM, and CM. These procedures have to cooperate in the form of structured procedures for the different transactions. The elements of a structured signaling procedure are

- Phase 1: Paging, channel request, assignment of a signaling channel
- Phase 2: Service request and collision resolution
- Phase 3: Authentication
- Phase 4: Activation of user data encryption
- Phase 5: Transaction phase
- Phase 6: Release and deallocation of the channel

Two examples of structured signaling procedures are presented in Figure 7.31 and Figure 7.32. They show the phases executed for the structured transaction, the terminating entities (MS, BSS, MSC), the respective message, and the logical channel used for the transport of the message. The first example (Figure 7.31) is a mobile-initiated call setup with OACSU − a traffic channel is only assigned after the subscriber of the called station is presented with the call request (ALERTING). The second example (Figure 7.32) shows a service change from voice to data and the modification of the selected data service. Such a modification could for example be the change of transmission rate. Finally the call is released and the traffic channel is deallocated.

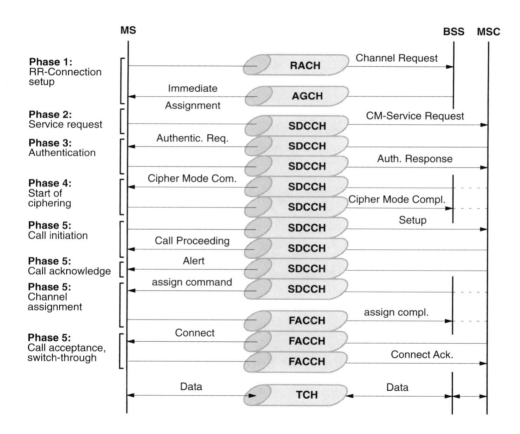

Figure 7.31: Mobile-initiated call setup with OACSU (*late assignment*)

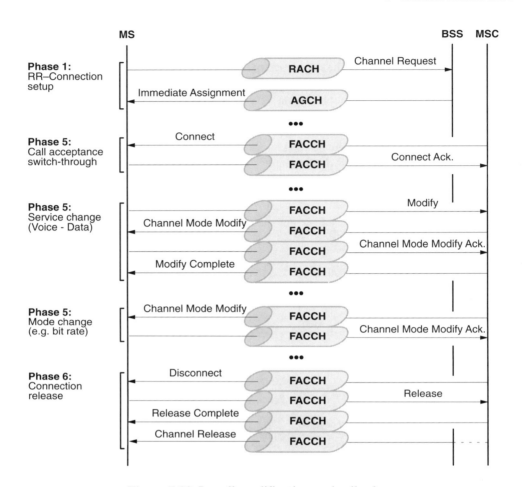

Figure 7.32: In-call modification and call release

7.4.7 Signaling Procedures for Supplementary Services

As can be seen in Figure 7.21, signaling messages to control *Supplementary Services* (SS) are coded with special protocol discriminators: 0011 for call related; 1011 for noncall related. A special set of signaling messages has been defined for their control (Table 7.7).

The category *CC Messages* (Table 7.6) consists of the subcategories *Call Information Phase* (message type MT=0x01tttt) and *Miscellaneous* (message type MT=0x11tttt). These two message categories are used in two categories of SS procedures: the *Separate Message Approach* and the *Common Information Element Procedure*. Whereas the *Separate Message Approach* uses its own messages (HOLD / RETRIEVE, Table 7.7) to activate specific functions, the functions of the *Common Information Element Procedure* are handled with a generic FACILITY message. Functions of the first category need synchronization between network and mobile station. The FACILITY category, however, is only used for supplementary services which do not require synchronization. This distinction becomes obvious in the examples of realized supplementary services, which will be presented in the following.

Table 7.7: CC messages for supplementary services

Category	Message	Direction	MT
Call Information Phase	Hold	N <-> MS	0x011000
	Hold Acknowledge	N <-> MS	0x011001
	Hold Reject	N <-> MS	0x011010
	Retrieve	N <-> MS	0x011100
	Retrieve Acknowledge	N <-> MS	0x011101
	Retrieve Reject	N <-> MS	0x011110
Miscellaneous	Facility	N <-> MS	0x111010
	Register	N <-> MS	0x111011

The messages of the *Separate Message Approach* can be used during the *Call Information Phase* to realize supplementary services like hold, callback, or call waiting. Figure 7.33 shows examples. A completely established call (call reference CR: 1 in Figure 7.33) can be put into the hold state from either one of the two partner entities.

To perform this supplementary service, it is initiated with a HOLD message. The MSC interrupts the connection and indicates with a HOLD message to the partner entity that the call is in the hold state. On each call segment this fact is acknowledged with a HOLD ACKNOWLEDGE message, which leads to both the requesting mobile station and the MSC to cut the traffic channel. The mobile station which caused the hold state to be entered can now establish another call (CR: 2 in Figure 7.33) or accept a call that may be coming in. Using another handshake HOLD/HOLD ACKNOWLEDGE, this call could be put into hold state too, and there could be switching between both held calls (brokering). For this purpose, a held call (CR: 1 in Figure 7.33) can be reactivated with a RETRIEVE message and reconnected to the call at each side of the traffic channel after the reactivation of the call has been acknowledged with a RETRIEVE ACKNOWLEDGE message.

These call-related signaling procedures modify the call state and define an extended state diagram with an auxiliary state. The participating calls all remain in the active state whereas the auxiliary state changes between hold and idle. In a two-dimensional state space, for example, call CR: 1 changes from (*Active, Idle*) through the state (*Active, Hold Request*) into the state (*Active, Call held*) and back through the state (*Active, Retrieve Request*).

If outgoing or incoming calls are barred (Figure 7.34), a call request is immediately refused by giving a RELEASE COMPLETE message with a reason in a FACILITY information element (BAOC, BAIC). A state change of the call in the extended state space does not occur. The assumption is, of course, that the calling or called subscriber has activated call barring. The MSC receiving the call request from the calling subscriber must verify the activation of this supplementary service. This requires an inquiry of the HLR of the calling subscriber (for BAOC) or the called subscriber (for BAIC), since the HLRs store the service profiles of the respective subscribers. In this case the HLR acts not only as a database but also as a participant in controlling intelligent network services.

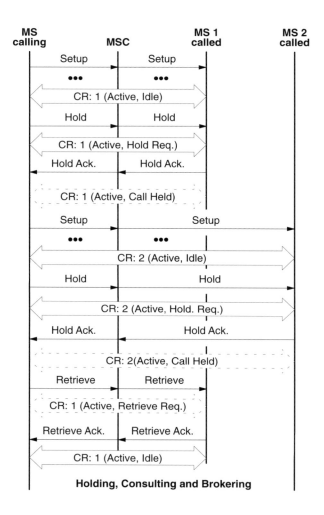

Figure 7.33: Call holding and associated procedures

Another call-related supplementary service uses the FACILITY message of the Common Information Element Procedure: *Call Forwarding Unconditional* (CFU); see Figure 7.34.
With this supplementary service, a regular call setup is performed, however, not to the
called subscriber but to the forwarding target selected when the service was activated (in
Figure 7.34 it is another MS). The calling subscriber is informed about the change of the
called number with a FACILITY message. Likewise the target of the forwarding is informed
with a FACILITY message that the incoming call is a forwarded call. In this case, there is no
necessity for a change in the extended state diagram nor is synchronization between network and mobile station required. It is only necessary that the involved mobile stations are
informed about the occurrence of forwarding. In this case, the target of the call forwarding
is also stored in the HLR of the subscriber who activated the service (the called MS in
Figure 7.34). Thus the call processing in the MSC of the called subscriber must be inter-

rupted and the HLR must be informed about the call request. If the called subscriber has activated unconditional call forwarding, the HLR returns the new call target to the MSC, which can continue call processing with the changed target number.

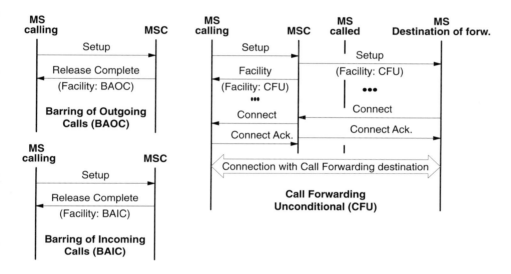

Figure 7.34: Barring and forwarding of calls

7.4.8   Realization of Short Message Services

The procedures for the transport of point-to-point short messages reside in the CM sublayer, also called the *Short Message Control Layer* (SM-CL) and in the *Short Message Relay Layer* (SM-RL) directly above. Accordingly, the protocol entities are called the *Short Message Control* entity (SMC) and the *Short Message Relay* entity (SMR). A complete established MM connection is needed for the transport of short messages, which again presumes an existing RR connection with LAPDm protection on an SDCCH or SACCH channel. To distinguish among these packet-switched user data connections, SMS messages are transported across SAPI=3 of the LAPDm entity.

An SMS transport PDU (SMS-SUBMIT or SMS-DELIVER, Figure 7.35) is transmitted with an RP-DATA message between MSC and MS using the *Short Message Relay Protocol* (SM-RP); see Section 7.3.2. Correct reception is acknowledged with an RP-ACK message either from the SMS service center (mobile-initiated SMS transfer) or from the MS (mobile-terminated SMS transfer).

For the transfer of short messages between SMR entities in MS and MSC, the CM sublayer provides a service to the SM-RL layer above. The SMR entity requests this service for the transfer of RP-DATA or RP-ACK (MNSMS-Establish-Request, Figure 7.36). Following the SMR service request, the SMC entity itself requests an MM connection on which it then transfers the short message inside a CP-DATA message. The appropriate service primitives between protocol layers are also shown for illustration purposes in Figure 7.36. The correct reception of CP-DATA is acknowledged with CP-ACK. In these SMC messages, one protocol data unit (PDU) transports a service data unit (SDU) from the SMR sublayer above. This SMC−SDU is the SMS relay message RP-DATA and its acknowledgment RP-ACK which are used to signal the transfer of short messages. The *Short Message Transport Layer* SM-TL above the SM-RL provides end-to-end transport of short messages between mobile station and *SMS Service Center* (SMS−SC).

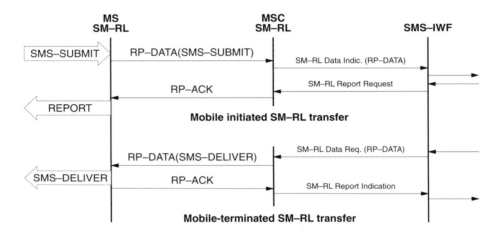

Figure 7.35: Short message transfer between SMR entities

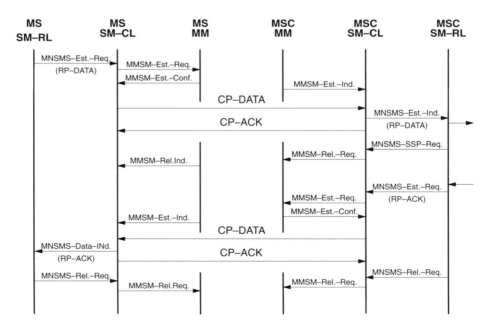

Figure 7.36: Short message transfer on the CM plane between MS and MSC

7.5   Signaling at the A and Abis Interfaces

Whereas the transport of user data between MSC and BSC occurs across standard connections of the fixed network with 64 kbit/s or 2048 kbit/s (or 1544 kbit/s), the transport of signaling messages between MSC and BSC runs over the SS#7 network. The MTP and SCCP parts of SS#7 are used for this purpose. A protocol function using the services of the SCCP is defined at the A interface. This is the *Base Station Application Part* (BSSAP), which is further subdivided into *Direct Transfer Application Part* (DTAP) and *Base Station System Management Part* (BSSMAP); see Figure 7.37. In addition, the *Base Station System Operation and Maintenance Part* (BSSOMAP) was introduced, which is needed for the transport of network management information from OMC via the MSC to the BSC.

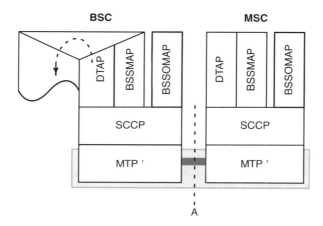

Figure 7.37: Protocols at the A interface between MSC and BSS

At the A interface, one can distinguish between two signaling message streams: one between MSC and MS and another between MSC and BSS. The messages to the mobile station (CM, MM) are passed on transparently through the BSS using the DTAP protocol part of SS#7. BSC and BTS do not interpret them. The SCCP protocol part provides a connection-oriented and a connectionless transfer service for signaling messages. For DTAP messages, only connection-oriented service is offered. The DTAP of the BSSAP uses one signaling connection for each active mobile station with one or more transactions per connection. A new connection is established each time when messages of a new transaction with a mobile station are to be transported between MSC and BSS.

Two cases of setting up a new SCCP connection are distinguished. First, in the case of location update and connection setup (outgoing or incoming), the BSS requests an SCCP connection after the channel request (access burst on RACH) from the mobile station has been satisfied with an SDCCH or TCH and after an LAPDm connection has been set up on the SDCCH or FACCH. The second situation for setting up an SCCP connection is a handover to an other BSS, in which case the MSC initiates the connection setup. Most of the signaling messages at the air interface (CM and MM, Table 7.5 and Table 7.6) are passed transparently through the BSS and packaged into DTAP−PDUs at the A interface, with the exception of some RR messages.

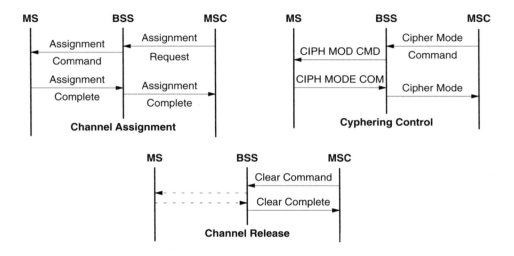

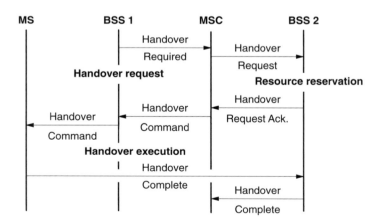

Figure 7.38: Examples of dedicated BSSMAP procedures

The BSSMAP implements two more kinds of signaling procedures between MSC and BSS, first those concerning one mobile station or single physical channels at the air interface, and second, global procedures for the control of all the resources of a BSS or cell. In the first case, the BSSMAP also uses connection-oriented SCCP services, whereas, in the second case, global procedures are performed with connectionless SCCP services. Among the BSSMAP procedures for a dedicated resource of the air interface are functions of resource management (channel assignment and release, start of ciphering) and of handover control (Figure 7.38 and Figure 7.39).

Figure 7.39: Dedicated BSSMAP procedures for internal handover

Among the global procedures of BSSMAP are paging, flow control to prevent overloading protocol processors or CCCH channels, closing and opening of channels, and parts of handover control (Figure 7.40).

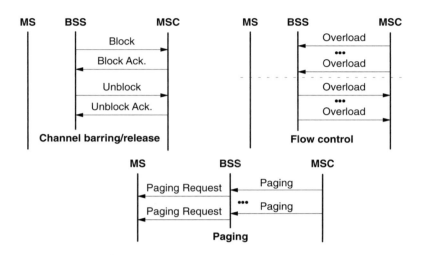

Figure 7.40: Examples of global BSSMAP procedures

The transmission layer at the Abis interface between BTS and BSC is usually realized as a primary multiplexed line with 2048 kbit/s (1544 kbit/s in North America) or 64 kbit/s. This may include one physical connection per BTS or one for each connection between TRX / BCF module and BTS (Figure 7.5). On these digital paths, traffic or signaling channels of 16 or 64 kbit/s are established. The Layer 2 protocol at the Abis interface is LAPDm, whose *Terminal Equipment Identifier* (TEI) is used to address the TRX and/or BCF of a BTS (Figure 7.41).

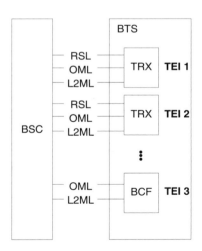

Figure 7.41: Logical connections at Layer 2 of the Abis interface

Several LAPDm connections are established for each TEI: the *Radio Signaling Link* (RSL), SAPI=0; the *Operation and Maintenance Link* (OML), SAPI=62; and the *Layer 2 Management Link* (L2ML), SAPI=63. Traffic management is handled on the RSL, operation and maintenance on the OML, and management messages of Layer 2 are sent on the L2ML to the TRX or BCF. The RSL is the most important of these three links for the control of radio resources and connections for communication between MS and network. Two types of messages are distinguished on this signaling link: transparent and nontransparent messages (Figure 7.42). Whereas the BTS passes transparent messages on from/to the LAPDm entity of an MS without interpreting or changing them, nontransparent messages are exchanged between BTS and BSC.

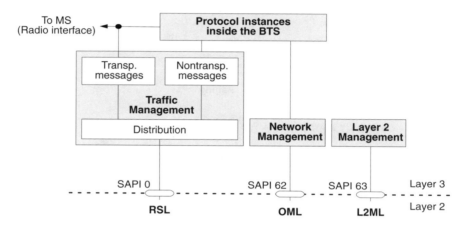

Figure 7.42: Protocol Layer 3 of the BTS at the Abis interface (BTSM)

In addition, one distinguishes between four groups of *Traffic Management* messages of the BTS:

- *Radio Link Layer Management*: procedures to establish, modify and release connections of the link layer (LAPDm) to the mobile station at the air interface Um.
- *Dedicated Channel Management*: procedures to start ciphering, transfer of channel measurement reports of an MS, transmitter power control of MS and BTS, handover detection, and modification of a dedicated channel of the BTS for a specific MS which can then receive the channel in an other message (assign, handover command).
- *Common Channel Management*: procedures for transferring channel requests from MS (received on RACH), start of paging calls, measurement and transfer of CCCH traffic load measurements, modification of BCCH broadcast information, channel assignments to the MS (on the AGCH), and transmission of *Cell Broadcast Short Messages* (SMSCB).
- *TRX Management*: procedures for the transfer of measurements of free traffic channels of a TRX to the BSC or for flow control in case of overloaded TRX processors or overload on the downlink CCCH/ACCH.

In this way, all the RR functions in the BTS can be controlled. The majority of RR messages (Table 7.4) are passed on transparently and do not terminate in the BTS. These messages are transported between BTS and BSC (Figure 7.43) in special messages (DATA REQUEST / INDICATION) packaged into LAPDm frames (Layer 2 at the radio interface).

Figure 7.43: Transfer of transparent signaling messages

All protocol messages received by the BTS on the uplink from the MS in LAPDm I/UI frames, except for channel measurement reports of the MS, are passed on as transparent messages in a DATA INDICATION.

Except for the link protocol LAPDm which is completely implemented in the BTS, there are some functions which are also handled by the BTS, and the pertinent messages from or to the MS are transformed by the BTS into the appropriate RR messages. This includes channel assignment, ciphering, assembly of channel measurements from MS and TRX, and their transfer to the BSC (possibly with processing in the BTS), power control commands from the BTS for the MS, and channel requests from the MS (on the RACH) as well as channel assignments (Figure 7.44). Thus four of the RR messages on the downlink direction to the MS (Table 7.4) cannot be treated as transparent messages: CIPHERING MODE COMMAND, PAGING REQUEST, SYSTEM INFORMATION and the three IMMEDIATE ASSIGN messages. All the other RR messages to the MS are sent transparently within a DATA REQUEST to the BTS.

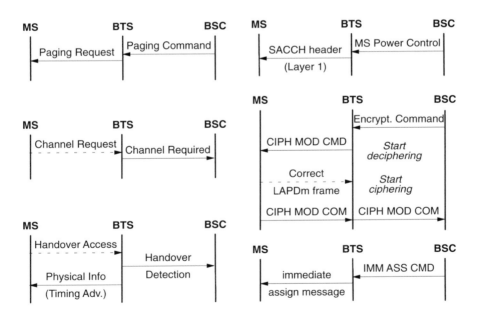

Figure 7.44: Examples of nontransparent signaling between BTS and BSC

Figure 7.45 shows the format of a BTSM message (Layer 3 between BSC and BTS). Transparent and nontransparent messages are distinguished with a *Message Discriminator* in the first octet. For this purpose, the T-bit (bit 1 of octet 1) is set to logical 1 for messages which the BTS is supposed to handle transparently or has recognized as transparent. Bits 2 to 5 serve to assign the messages to one of the four groups defined on the *Radio Signaling Link* (RSL). Including the *Message Type* (MT) defines the message completely (Figure 7.45). The remainder of the BTSM message contains mandatory and optional *Information Elements* (IEs) which have a fixed length of mostly two octets or which contain an additional length indicator in the case of variable length.

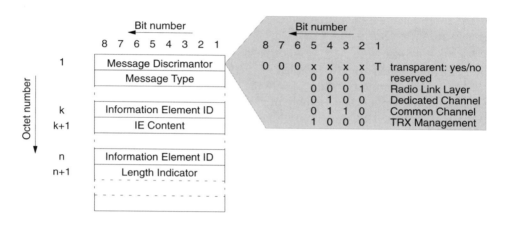

Figure 7.45: Format of BTSM-RSL protocol messages

7.6   Signaling at the User Interface

Another often neglected but nevertheless very important interface in a mobile system is the user interface of the mobile station equipment. This *Man–Machine Interface* (MMI) can be realized freely and therefore in many different ways by the mobile equipment manufacturers. In order to keep a set of standardized service control functions in spite of this variety, the MMI commands have been introduced. These MMI commands define procedures mainly for the control of basic and supplementary services. The control procedures are constructed around the input of command token strings which are delineated and formatted with the tokens * and #. In order to avoid a user having to learn and memorize a certain number of service control procedures before being able to use the mobile phone, a small set of basic required commands for the MMI interface has been defined; this is the *Basic Public MMI* which must be satisfied by all mobile stations.

The specification *Basic Public MMI* outlines the basic functions which must be implemented as a minimum at the MMI of a mobile station. This includes the arrangement of a 12-key keyboard with the numbers 0 through 9 and the keys for * and # as well as SEND and END keys, which also serve to initiate a desired call or accept or terminate a call. Some basic operational sequences for making or taking a call are also defined. These requirements are so general that they can be easily satisfied by all mobile equipment.

Table 7.8: Input format of some MMI commands

Function	MMI procedure
Activate	* nn(n)*Si#
Deactivate	#nn(n)*Si#
Status enquiry	*#nn(n)*Si#
Registration	**nn(n)*Si#
Delete	##nn(n)*Si#

The MMI commands for the control of supplementary services and the enquiry and configuration of parameters are much more extensive. Using a set of MMI commands which are uniform for all mobile stations allows control functions to be performed which are often hidden in equipment-specific user guidance menus. For certain functional areas, a mobile station can thus be operated in a manufacturer-independent way, if one forgoes the sometimes very comfortable possibilities of user-guiding menus and instead learns the control sequences for the respective functions. These sequences are mapped onto the respective signaling procedures within the mobile station.

An MMI command is always constructed according to the same pattern. Five basic formats are distinguished (Table 7.8), which all start with a combination of the tokens * and #: activation (*), deactivation (#), status inquiry (*#), registration (**) and cancellation (##).

In addition, the MMI command must contain an MMI service code of two or three tokens, which selects the function to be performed. In certain cases, the MMI procedure requires additional arguments or parameters, which are separated by *, as *Supplementary Information* (Si). The MMI command is always terminated with # and may also require depressing the SEND key, if the command is not executed locally within the mobile station but must be transmitted to the network. Table 7.9 contains some basic examples of MMI commands, e.g. the inquiry for the IMEI of the MS (*#06#) or the change of the PIN (**04*old_PIN*new_PIN*new_PIN#) used to protect the SIM card against misuse. This example also shows how supplementary information is embedded in the command.

Table 7.9: Some basic MMI commands

Function	MMI-Procedure
Mobile phone IMEI enquiry	*#06#
Change password for call barring	**03*330*old_PWD*new_PWD*new_PWD#
Change PIN in SIM	**04*old_PIN*new_PIN*new_PIN#
Select SIM number storage	n(n)(n)#

With MMI commands, it is also possible to configure and use *Supplementary Services* (see Section 4.3). For this purpose, each supplementary service is designated with a two- or three-digit MMI service code to select the respective supplementary service (Table 7.10). In some cases, supplementary information is mandatory for the activation of the service, e.g. one needs the target *Destination Number* (DN) for the call forwarding functions, or the *Activation Password* (PW) for the supplementary service of barring incoming or outgoing calls (*Sia* in Table 7.10).

Table 7.10: MMI service codes for supplementary services

Abbr.	Service	MMI code	Sia	Sib
	All Call Forwarding, only for (de)activation	002	–	–
	All conditional Call Forwarding (not CFU), only for (de-)activation	004	–	–
CFU	Call Forwarding Unconditional	21	DN	BS
CFB	Call Forwarding on Mobile Subscriber Busy	67	DN	BS
CFNRy	Call Forwarding on No Reply	61	DN	BS
CFNRc	Call Forwarding on Mobile Subscriber Not Reachable	62	DN	BS
	All Call Barring (only for deactivation)	330	PW	BS
BAOC	Barring of All Outgoing Calls	33	PW	BS
BOIC	Barring of Outgoing International Calls	331	PW	BS
BOIC-exHC	Barring of Outgoing International Calls Except those to Home PLMN	332	PW	BS
BAIC	Barring of All Incoming Calls	35	PW	BS
BIC-Roam	Barring of Incoming Calls when Roaming Outside the Home PLMN	351	PW	BS
CLIP	Calling Line Identification Presentation	30	–	BS
CLIR	Calling Line Identification Restriction	31	–	BS
CW	Call Waiting	43	–	BS
COLP	Connected Line Identification Presentation	76	–	BS
COLR	Connected Line Identification Restriction	77	–	BS

BS=Basic Service (see Table 7.11)   DN=Destination Number   PW=Password

The example of unconditional call forwarding also illustrates the difference between registration and activation of a service. With the command **21*call_number # the forwarding function is registered, the target number configured, and the unconditional forwarding activated. Later, the unconditional forwarding can be deactivated any time with #21# and reactivated with *21#. The target number *call_number* remains stored, unless it is canceled with the command ##21#. After cancellation, if call forwarding is desired again, it must first be registered again using the **21... command. For some basic services, characteristics can also be activated selectively. The MMI command can contain a second parameter field with supplementary information (*Sib*, Table 7.10) which is again delineated with *. This field contains the service code BS of the *Basic Service* for which the supplementary service is to become effective.

An overview of MMI codes for basic services is given in Table 7.11. For example, one can bar incoming calls except short messages with the command **35*PW*18#, or one can forward incoming fax calls to the number *fax_number* with the command **21*fax_number *13# (the other teleservices remain unaffected).

Table 7.11: MMI codes for basic services

Category	Service	MMI code BS
Telematic service	All telematic services	10
	Telephone	11
	All data services	12
	Facsimile	13
	Videotex	14
	Teletext	15
	Short Message Service (SMS)	16
	All data services except SMS	18
	All telematic services except SMS	19
Bearer service	All bearer services	20
	All asynchronous services	21
	All synchronous services	22
	All connection-oriented synchronous data services	24
	All connection-oriented asynchronous data services	25
	All packet-oriented synchronous data services	26
	All PAD-access services	27

8 Roaming and Switching

8.1 Mobile Application Part Interfaces

The main benefit for the mobile subscribers that the international standardization of GSM has brought is that they can move freely not only within their home networks but also in international GSM networks and that at the same time they can even get access to the special services they subscribed to at home − provided there are agreements between the operators. The functions needed for this free roaming are called *roaming* or *mobility functions*. They rely mostly on the GSM-specific extension of the *Signaling System Number 7* (SS#7). The *Mobile Application Part* (MAP) procedures relevant for roaming are first the *Location Registration/Update, IMSI Attach/Detach*, requesting subscriber data for call setup, and paging. In addition, the MAP contains functions and procedures for the control of *Supplementary Services* and handover, for subscriber management, for IMEI management, for authentication and identification management, as well as for the user data transport of the *Short Message Service*. MAP entities for roaming services reside in the MSC, HLR, and VLR. The corresponding MAP interfaces are defined as B (MSC-VLR), C (MSC-HLR), D (HLR-VLR), E (MSC-MSC), and G (VLR-VLR) (Figure 3.9). At the subscriber interface, the MAP functions correspond to the functions of *Mobility Management*, i.e. the MM messages and procedures of the Um interface are translated into the MAP protocols in the MSC.

The most important functions of GSM *Mobility Management* are *Location Registration* with the PLMN and *Location Updating* to report the current location of an MS, as well as the identification and authentication of subscribers. These actions are closely interrelated. During registration into a GSM network, during the location updating procedure, and also during the setup of a connection, the identity of a mobile subscriber must be determined and verified (authentication).

The mobility management data are the foundation for creating the functions needed for routing and switching of user connections and for the associated services. For example, they are requested for routing an incoming call to the current MSC or for localizing an MS before paging is started. In addition to mobility data management, information about the configuration of supplementary services is requested or changed, e.g. the currently valid target number for unconditional call forwarding in the HLR or VLR registers.

8.2   Location Registration and Location Update

Before a mobile station can be called or gets access to services, the subscriber has to regis-ter with the mobile network (PLMN). This is usually the home network where the subscri-ber has a service contract. However, the subscriber can equally register with a foreign net-work provider in whose service area he or she is currently visiting, provided there is a roaming agreement between the two network operators.

Registration is only required if there is a change of networks, and therefore a VLR of the current network has not yet issued a TMSI to the subscriber. This means the subscriber has to report to the current network with his IMSI and receives a new TMSI by executing a *Location Registration* procedure. This TMSI is stored by the MS in its nonvolatile SIM storage, such that even after a powerdown and subsequent power-up only a normal *Loca-tion Updating* procedure is required.

The sequence of operations for registration is presented schematically in Figure 8.1. After a subscriber has requested registration at its current location by sending a LOCATION UP-DATE REQUEST with its IMSI and the current location area (LA), first the MSC instructs the VLR with a MAP message UPDATE LOCATION AREA to register the MS with its current LAI. In order for this registration to be valid, the identity of the subscriber has to be checked first, i.e. the authentication procedure is executed. For this purpose, the authenti-cation parameters have to be requested from the AUC through the HLR. The precalcu-lated sets of security parameters (Kc, RAND, SRES) are usually not transmitted individu-ally to the respective VLR. In most cases, several complete sets are kept at hand for several authentications. Each set of parameters, however, can only be used once, i.e. the VLR must continually update its supply of security parameters (AUTHENTICATION PARAMETER REQUEST).

After successful authentication (see Section 6.3.2), the subscriber is assigned a new MSRN, which is stored with the LAI in the HLR, and a new TMSI is also reserved for this subscriber; this is *TMSI Reallocation* (see Figure 7.25). To encrypt the user data, the base station needs the ciphering key Kc, which it receives from the VLR by way of the MSC with the command START CIPHERING. After ciphering of the user data has begun, the TMSI is sent in encrypted form to the mobile station. Simultaneously with the TMSI assignment, the correct and successful registration into the PLMN is acknowledged (LOCATION UPDATE ACCEPT). Finally, the mobile station acknowledges the correct reception of the TMSI (TMSI REALLOCATION COMPLETE, see Figure 7.26).

While the location information is being updated, the VLR is obtaining additional infor-mation about the subscriber, e.g. the MS category or configuration parameters for supple-mentary services. For this purpose, the *Insert Subscriber Data Procedure* is defined (INSERT SUBSCRIBER DATA message in Figure 8.1). It is used for registration or location updating in the HLR to transmit the current data of the subscriber profile to the VLR. In general, this MAP procedure can always be used when the profile parameters are changed, e.g. if the subscriber reconfigures a supplementary service such as unconditional forwarding. The changes are communicated immediately to the VLR with the *Insert Subscriber Data Procedure.*

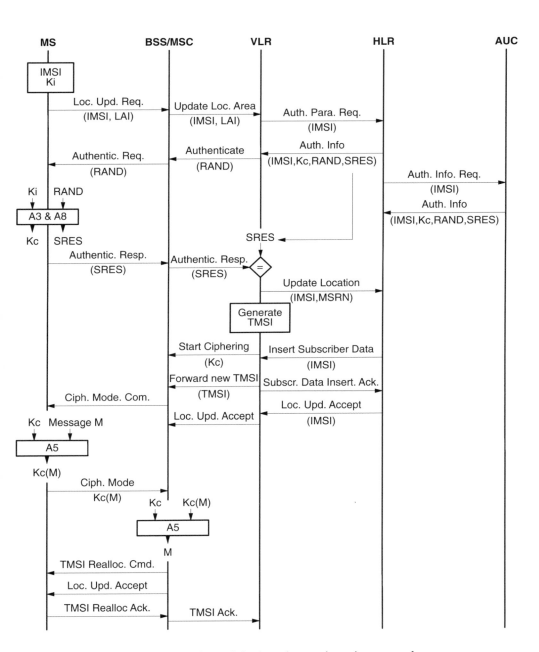

Figure 8.1: Overview of the location registration procedure

The location update procedure is executed, if the mobile station recognizes by reading the LAI broadcast on the BCCH that it is in a new location area, which leads to updating the location information in the HLR record. Alternatively, the location update can also occur periodically, independent of the current location. For this purpose, a time interval value is broadcast on the BCCH, which prescribes the time between location updates. The main objective of this location update is to know the current location for incoming calls or short messages, so that the call or message can be directed to the current location of the mobile station. The difference between the location update procedure and the location registration procedure is that in the first case the mobile station has already been assigned a TMSI. The TMSI is unique only in connection with an LAI, and both are kept together in the nonvolatile storage of the SIM card. With a valid TMSI, the MS also keeps a current ciphering key Kc for encryption of user data (Figure 8.2), although this key is renewed during the location update procedure. This key is recalculated by the MS based on the random number RAND used for authentication, whereas on the network side it is calculated in the AUC and made available in the VLR.

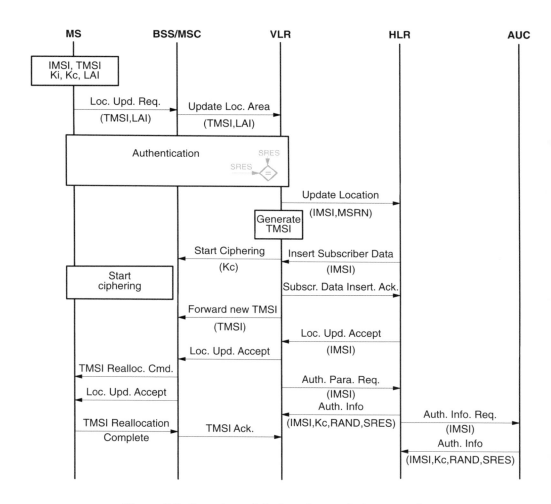

Figure 8.2: Overview of the location updating procedure

Corresponding to the location update procedure, there is an MM procedure at the air interface of the MM-category *Specific*. Besides the location updating proper, it contains three blocks which are realized at the air interface by three procedures of the category

Common (see Figure 7.26): the identification of the subscriber, the authentication, and the start of ciphering on the radio channel. In the course of location updating, the mobile station also receives a new TMSI, and the current location is updated in the HLR. Figure 8.2 illustrates the standard case of a location update. The MS has entered a new LA, or the timer for periodic location updating has expired, and the MS requests to update its location information. It is assumed that the new LA still belongs to the same VLR as the previous one, so only a new TMSI needs to be assigned. This is the most frequent case. But if its not quite so crucial to keep the subscriber identity confidential, it is possible to avoid assigning a new TMSI. In this case, only the location information is updated in the HLR / VLR.

The new TMSI is transmitted to the MS in enciphered form together with the acknowledgment of the successful location update. The location update is complete after acknowledgment by the mobile station. After execution of the authentication, the VLR can complete its database and replace the "consumed" 3-tuple (RAND, SRES, Kc) by an other one requested from the HLR / AUC.

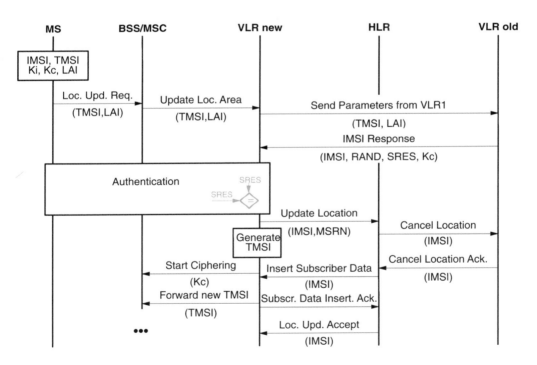

Figure 8.3: Location update after changing the VLR area

If location change involves both LA and VLR, the location update procedure is somewhat more complicated (Figure 8.3). In this case, the new VLR has to request the identification and security data for the MS from the old VLR and store them locally. Only in emergency cases, if the old VLR cannot be determined from the old LAI or if the old TMSI is not known in the VLR, the new VLR may request the IMSI directly from the MS (identification procedure).

Only after a mobile station has been identified through the IMSI from the old VLR and after the security parameters are available in the new VLR, is it possible for the mobile station to be authenticated and registered in the new VLR, for a new TMSI to be assigned, and for the location information in the HLR to be actualized. After successful registration in the new VLR (LOCATION UPDATE ACCEPT) the HLR instructs the old VLR to cancel the invalid location data in the old VLR (CANCEL LOCATION).

In the examples shown (Figures 8.1, 8.2, and 8.3), the location information is stored as MSRN in the HLR. The MSRN contains the routing information for incoming calls and this information is used to route incoming calls to the current MSC. In this case, the HLR receives the routing information already at the time of the location update. Alternatively, at location update time the HLR may just store the current MSC and/or VLR number in connection with an LMSI, such that routing information is only determined at the time of an incoming call.

8.3   Connection Establishment and Termination

8.3.1   Routing calls to Mobile Stations

The number dialed to reach a mobile subscriber (MSISDN) contains no information at all about the current location of the subscriber. In order to establish a complete connection to a mobile subscriber, however, one must determine the current location and the locally responsible switch (MSC). In order to be able to route the call to this switch, the routing address to this subscriber (MSRN) has to be obtained. This routing address is assigned temporarily to a subscriber by its currently associated VLR. At the arrival of a call at the GMSC, the HLR is the only entity in the GSM network which can supply this information, and therefore it must be interrogated for each connection setup to a mobile subscriber. The principal sequence of operations for routing to a mobile subscriber is shown in Figure 8.4. An ISDN switch recognizes from the MSISDN that the called subscriber is a mobile subscriber, and therefore can forward the call to the GMSC of the subscriber's home PLMN based on the CC and NDC in the MSISDN (1). This GMSC can now request the current routing address (MSRN) for the mobile subscriber from the HLR using the MAP (2,3). By way of the MSRN the call is forwarded to the local MSC (4), which determines the TMSI of the subscriber (5,6) and initiates the paging procedure in the relevant location area (7). After the mobile station has responded to the paging call (8), the connection can be switched through.

Several variants for determining the route and interrogating the HLR exist, depending on how the MSRN was assigned and stored, whether the call is national or international, and depending on the capabilities of the associated switching centers.

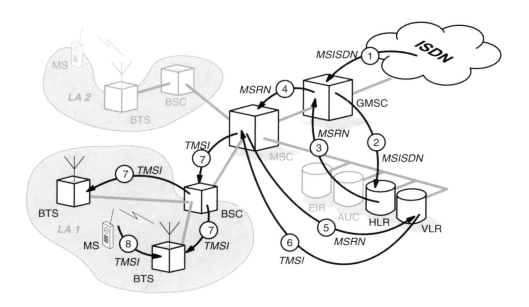

Figure 8.4: Principle of routing calls to mobile subscribers

8.3.1.1   Effect of the MSRN Assignment on Routing

There are two ways to obtain the MSRN:

- obtaining the MSRN at location update
- obtaining the MSRN on a per call basis

For the first variant, an MSRN for the mobile station is assigned at the time of each location update which is stored in the HLR. This way the HLR is in a position to supply immediately the routing information needed to switch a call through to the local MSC.

The second variant requires that the HLR has at least an identification for the currently responsible VLR. In this case, when routing information is requested from the HLR, the HLR first has to obtain the MSRN from the VLR. This MSRN is assigned on a per call basis, i.e. each call involves a new MSRN assignment.

8.3.1.2   Placement of the Protocol Entities for HLR Interrogation

Depending on the capabilities of the associated switches and the called target (national or international MSISDN), there are different routing procedures. In general, the local switching center analyzes the MSISDN. Due to the NDC, this analysis of the MSISDN allows the separation of the mobile traffic from other traffic. The case that mobile call numbers are integrated into the numbering plan of the fixed network is currently not provided.

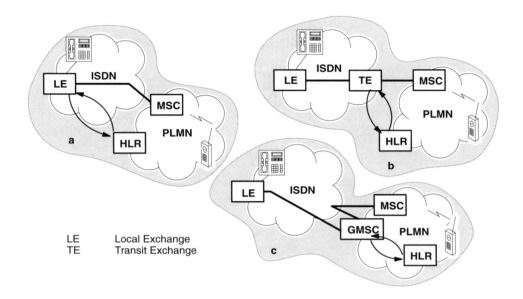

Figure 8.5: Routing variants for national MSISDN

In the case of a national number, the local exchange recognizes from the NDC that the number is a mobile ISDN number. The fixed network and home PLMN of the called subscriber reside in the same country. In the ideal case, the local switch can interrogate the HLR responsible for this MSISDN (HLR in the home PLMN of the subscriber) and obtain the routing information (Figure 8.5a). The connection can then be switched through via fixed connections of the ISDN directly to the MSC.

If the local exchange does not have the required protocol intelligence for the interrogation of the HLR, the connection can be passed on preliminarily to a transit exchange, which then assumes the HLR interrogation and routing determination to the current MSC (Figure 8.5b). If the fixed network is not at all capable of performing an HLR interrogation, the connection has to be directed through a gateway MSC (GMSC). This GMSC connects through to the current MSC (Figure 8.5c). For all three cases, the mobile station could also reside in a foreign PLMN (roaming); the connection is then made through international lines to the current MSC after interrogating the HLR of the home PLMN.

In the case of an international call number, the local exchange recognizes only the international country code (CC) and directs the call to an *International Switching Center* (ISC). Then the ISC can recognize the NDC of the mobile network and process the call accordingly. Figure 8.6 and Figure 8.7 show examples for the processing of routing information. An international call to a mobile subscriber involves at least three networks: the country from which the call originates; the country with the home PLMN of the subscriber, *Home PLMN* (H-PLMN); and the country in which the mobile subscriber is currently roaming, *Visited PLMN* (V-PLMN). The traffic between countries is routed through *International Switching Centers* (ISCs). Depending on the capabilities of the ISC, there are several routing variants for international calls to mobile subscribers. The difference is determined by the entity that performs the HLR interrogation, resulting in differently occupied line capacities.

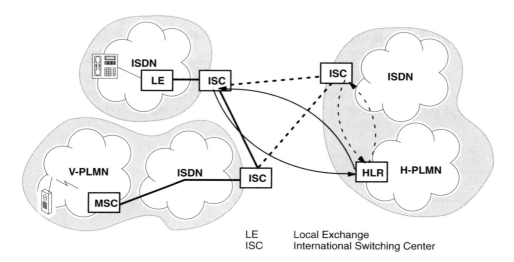

Figure 8.6: Routing for international MSISDN (HLR interrogation from ISC)

If the ISC performs the HLR interrogation, the routing to the current MSC is performed either by the ISC of the originating call or by the ISC of the mobile subscriber's H-PLMN (Figure 8.6). If no ISC can process the routing, again a GMSC has to get involved, either a GMSC in the country where the call originates or the GMSC of the H-PLMN (Figure 8.7).

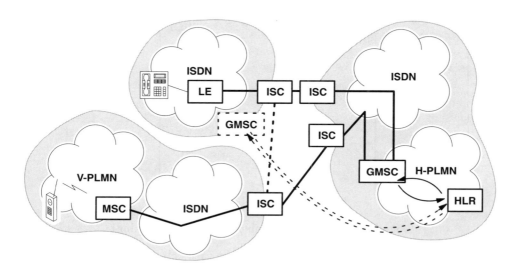

Figure 8.7: Routing through GMSC for international MSISDN

For the routing procedures explained here, it does not matter which kind of subscriber is calling, i.e. the subscriber may be in the fixed network or in the mobile network. However, for calls from mobile subscribers, the HLR interrogation is usually performed at the local exchange (MSC).

8.3.2 Call Establishment and Corresponding MAP Procedures

Call establishment in GSM at the air interface is similar to ISDN call establishment at the user network interface (Q.931) [22]. The procedure is supplemented by several functions: *Random Access* to establish a signaling channel (SDCCH) for call setup signaling, the authentication part, the start of ciphering, and the assignment of a radio channel.

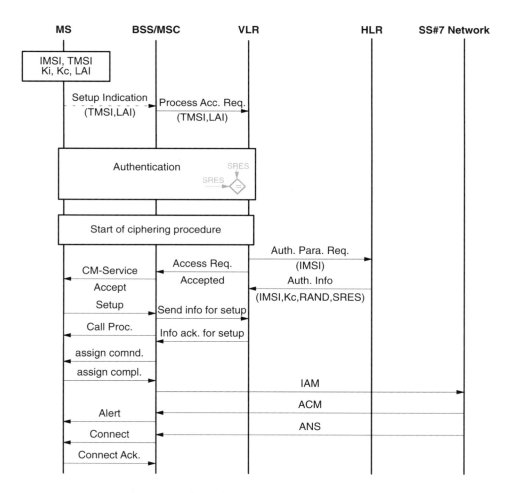

Figure 8.8: Overview of connection setup (outgoing)

The establishment of a connection always contains a verification of user identity (authentication) independent of whether it is a *mobile-originated call setup* or a *mobile-terminated call setup*. The authentication is performed in the same way as for location updating. The

VLR supplements its database entry for this mobile station with a set of security data, which replaces the "consumed" 3-tuple (RAND, SRES, Kc). After successful authentication, the ciphering process for the encryption of user data is started.

For outgoing connection setup (Figure 8.8), first the mobile station announces its connection request to the MSC with a SETUP INDICATION message, which is a pseudomessage. It is generated between the MM entity of the MSC and the MAP entity, when the MSC receives the message CM-SERVICE REQUEST from the MS, which indicates in this way the request for an MM connection (see Figure 7.27). Then the MSC signals to the VLR that the mobile station identified by the temporary TMSI in the location area LAI has requested service access (PROCESS ACCESS REQUEST) which is an implicit request for a random number RAND from the VLR, to be able to start the authentication of the MS. This random number is transmitted to the mobile station, its response with authentication result SRES is returned to the VLR, which now examines the authenticity of the mobile station's identity (compare authentication at registration, Figure 8.1).

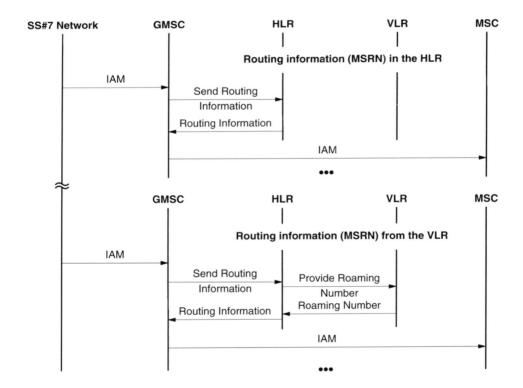

Figure 8.9: Interrogation of routing information for incoming call

After successful authentication, the ciphering process is started on the air interface, and this way the MM connection between MS and MSC has been completely established (CM-SERVICE ACCEPT). Subsequently, all signaling messages can be sent in encrypted form. Only now the MS reports the desired calling target. While the MS is informed with a CONNECT ACKNOWLEDGE message that processing of its connection request has been started, the

MSC reserves a channel for the conversation and assigns it to the MS (ASSIGN). The connection request is signaled to the remote network exchange through the signaling system SS#7 with the *ISDN User Part* (ISUP) message IAM [22]. When the remote exchange answers (ACM), the delivery of the call can be indicated to the mobile station (ALERT). Finally, when the called partner goes off-hook, the connection can be switched through (CONNECT, ANS, CONNECT ACKNOWLEDGE).

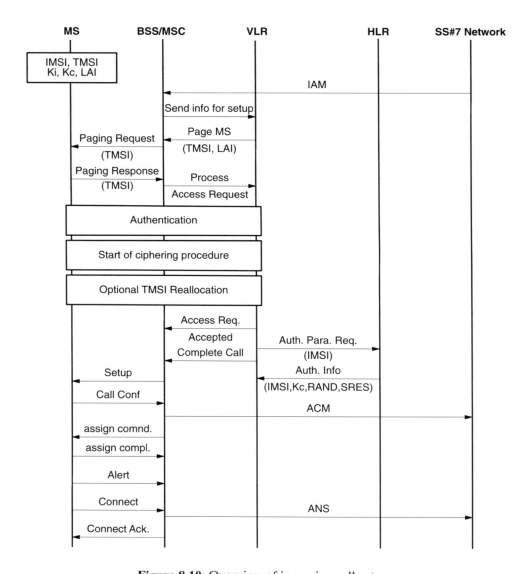

Figure 8.10: Overview of incoming call setup

For incoming connection setup, it is necessary to determine the exact location of an MS in order to route the call to the currently responsible MSC. A call to a mobile station is therefore always routed to an entity which is able to interrogate the HLR for temporary routing information and to use it to forward the call. Usually, this entity is a GMSC of the home PLMN of the MS (see Section 8.3.1.2). Through this HLR interrogation, the GMSC obtains the current MSRN of the mobile station and forwards it to the current MSC (Figure 8.9).

Depending on whether the MSRN is stored in the HLR or first has to be obtained from the VLR, two variants of the HLR interrogation exist. In the first case, the interrogated HLR can supply the MSRN immediately (ROUTING INFORMATION). In the second case, the HLR has only received and stored the current VLR address during location update. Therefore, the HLR first has to request the current routing information from the VLR before the call can be switched through to the local MSC.

Call processing is interrupted again in the local MSC in order to determine the exact location of the mobile station within the MSC area (SEND INFO FOR SETUP, Figure 8.10). The current LAI is stored in the location registers, but an LA can comprise several cells. Therefore, a broadcast (paging call) in all cells of the LA is used to determine the exact location, i.e. cell, of the MS. Paging is initiated from the VLR using the MAP (PAGE MS) and transformed by the MSC into the paging procedure at the air interface. When an MS receives a paging call, it responds directly and thus allows determination of the current cell.

Thereafter, the VLR instructs the MSC to authenticate the MS and to start ciphering on the signaling channel. Optionally, the VLR can execute a reallocation of the TMSI (TMSI reallocation procedure) during call setup. Only at this point, after the network internal connection has been established (see Section 7.4.4), the connection setup proper can be processed (command COMPLETE CALL from VLR to MSC). The MS is told about the connection request with a SETUP message, and after answering CALL COMPLETE it receives a channel. After ringing (ALERT) and going off-hook, the connection is switched through (CONNECT, CONNECT ACKNOWLEDGE), and this fact is also signaled to the remote exchange (ACM, ANS).

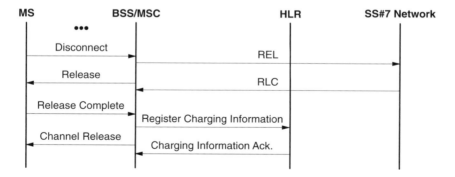

Figure 8.11: Mobile-initiated call termination and storing of charging information

8.3.3   Call Termination

At the air interface, a given call can be terminated either by the mobile equipment or by the network. The taking down of the connection is initiated at the Um interface by means of the CC messages DISCONNECT, RELEASE and RELEASE COMPLETE. This is followed by an explicit release of occupied radio resources (CHANNEL RELEASE). On the network side, the connection between the involved switching centers (MSC, etc.) is terminated using the ISUP messages REL and RLC in the SS#7 network (Figure 8.11).

After taking down of the connection, information about charges (CHARGING INFORMATION) is stored in the VLR or HLR using the MAP. This charging data can also be required for an incoming call, e.g. if roaming charges are due because the called subscriber is not in his or her home PLMN.

8.3.4 MAP Procedures and Routing for Short Messages

A connectionless relay protocol has been defined for the transport of short messages (see Section 7.4.8) at the air interface, which has a counterpart in the network in a store-and-forward operation for short messages. This forwarding of transport PDUs of the SMS uses MAP procedures. For an incoming short message which arrives from the *Short Message Service Center* (SMS-SC) at a *Short Message Gateway MSC* (SMS-GMSC), the exact location of the MS is the first item that needs to be determined just as for an incoming call. The current MSC of the MS is first obtained with an HLR interrogation (SHORT MESSAGE ROUTING INFORMATION, Figure 8.12a). The short message is then passed to this MSC (FORWARD SHORT MESSAGE) and is locally delivered after paging and SMS connection setup. Success or failure are reported to the SMS-GMSC in another MAP message (FORWARD ACKNOWLEDGMENT/ERROR INDICATION) which then informs the service center.

In the reverse case, for an outgoing short message, no routing interrogation is needed, since the SMS-GMSC is known to all MSC, so the message can be passed immediately to the SMS-GMSC (Figure 8.12b).

8.4   Handover

8.4.1 Overview

Handover is the transfer of an existing voice connection to a new base station. There are different reasons for the handover to become necessary. In GSM, a handover decision is made by the network, not the mobile station, and it is based on BSS criteria (received signal level, channel quality, distance between MS and BTS) and on network operation criteria (e.g. current traffic load of the cell and ongoing maintenance work).

The functions for preparation of handover are part of the *Radio Subsystem Link Control*. Above all, this includes the measurement of the channel. Periodically, a mobile station checks the signal field strength of its current downlinks as well as those of the neighboring base stations, including their BSICs. The MS sends measurement reports to its current base station (quality monitoring); see Section 5.5.1. On the network side, the signal quality of the uplink is monitored, the measurement reports are evaluated, and handover decisions are made.

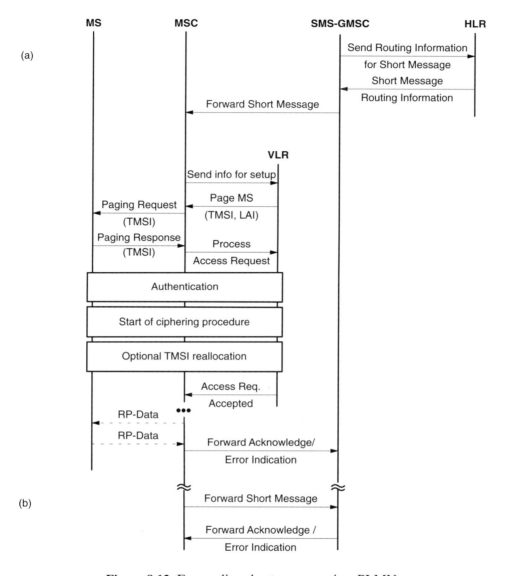

Figure 8.12: Forwarding short messages in a PLMN

As a matter of principle, handovers are only performed between base stations of the same PLMN. Handovers between BSS in different networks are not allowed. Two kinds of handover are distinguished (Figure 8.13):

- *Intracell Handover*: for administrative reasons or because of channel quality (channel-selective interferences), the mobile station is assigned a new channel within the same cell. This decision is made locally by the *Radio Resource Management* (RR) of the BSS and is also executed within the BSS.

- *Intercell Handover*: the connection to an MS is transferred over the cell boundary to a new BTS. The decision about the time of handover is made by the RR protocol module of the network based on measurement data from MS and BSS. The MSC, however, can participate in the selection of the new cell or BTS. The intercell hand-over occurs most often when it is recognized from weak signal field strength and bad channel quality (high bit error ratio) that a mobile station is moving near the cell boundary. However, an intercell handover can also occur due to administrative rea-sons, say for traffic load balancing. The decision about such a *network-directed handover* is made by the MSC, which instructs the BSS to select candidates for such a handover.

Two cases need to be distinguished with regard to participation of network components in the handover, depending on whether the signaling sequences of a handover execution also involve an MSC. Since the RR module of the network resides in the BSC (see Figure 7.11), the BSS can perform the handover without participation of the MSC. Such handovers occur between cells which are controlled by the same BSC and are called inter-nal handover. They can be performed independently by the BSS; the MSC is only in-formed about the successful execution of internal handovers. All other handovers require participation of at least one MSC, or their BSSMAP and MAP parts, respectively. These handovers are known as external handovers.

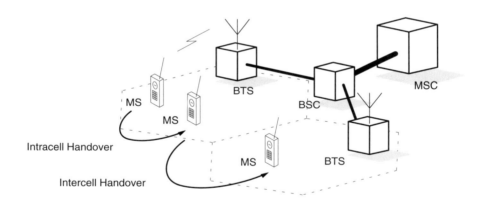

Figure 8.13: Intracell and Intercell handover

Participating MSCs can act in the role of MSC-A or MSC-B. MSC-A is the MSC which performed the initial connection setup, and it keeps the MSC-A role and complete control (anchor MSC) for the entire life of the connection. A handover is therefore in general the extension of the connection from the anchor MSC-A to another MSC (MSC-B). In this case, the mobile connection is passed from MSC-A to MSC-B with MSC-A keeping the ultimate control over the connection.

An example is presented in Figure 8.14. A mobile station occupies an active connection via BTS1 and moves into the next cell. This cell of BTS2 is controlled by the same BSC so that an internal handover is indicated. The connection is now carried from MSC-A over the BSC and the BTS2 to the mobile station; the connections of BTS1 (radio channel and ISDN channel between BTS and BSC) were taken down. As the mobile station moves on to the cell handled by BTS3, it enters a new BSS which requires an external handover. Be-sides, this BSS belongs to another MSC, which now has to play the role of MSC-B. Log-ically, the connection is extended from MSC-A to MSC-B and carried over the BSS to the mobile station. At the next change of the MSC, the connection element between MSC-A and MSC-B is taken down, and a connection to the new MSC from MSC-A is set up. Then the new MSC takes over the role of MSC-B.

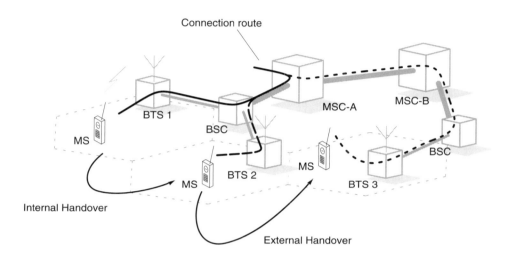

Figure 8.14: Internal and external handover

8.4.2   GSM Intra-MSC Handover

The basic structure for an external handover is the handover between two cells of the same MSC (Figure 8.15). The mobile station continually transmits measurement reports with channel monitoring data on its SACCH to the current base station (BSS 1). Based on these measurement results, the BSS decides when to perform a handover and requests this handover from the MSC (message HANDOVER REQUIRED). The respective measurement results can be transmitted in this message to the MSC, to enable its participation in the handover decision. The MSC causes the new BSS to prepare a channel for the handover, and frees the handover to the mobile station (HANDOVER COMMAND), as soon as the reservation is acknowledged by the new BSS. The mobile station now reports to the new BSS (HANDOVER ACCESS) and receives information about the physical channel properties. This includes synchronization data like the new timing advance value and also the new transmitter power level. Once the mobile station is able to occupy the channel successfully, it acknowledges this fact with a message HANDOVER COMPLETE. The resources of the old BSS can then be released.

8.4.3   Decision Algorithm for Handover Timing

The basis for processing a successful handover is a decision algorithm which uses measurement results from mobile and base station to identify possible other base stations as targets for handovers and which determines the optimal moment to execute the handover. The objective is to keep the number of handovers per cell change as small as possible. Ideally, there should not be more than one handover per cell change. In reality, this is often not achievable. When a mobile station leaves the radio range of a base station and enters one of a neighboring station, the radio conditions are often not very stable, so that several handovers must be executed before a stable state is reached. Simulation results in [5] and [6] give a mean value of about 1.5 to 5 handovers per cell change.

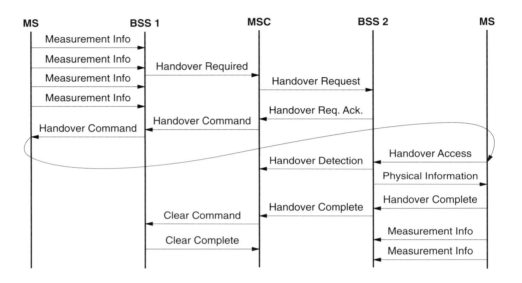

Figure 8.15: Principal signaling sequence for an intra-MSC handover

Since every handover incurs not only increased traffic load for the signaling and transport system but also reductions in speech quality, the importance of a well-dimensioned handover decision algorithm is obvious, an algorithm which also takes into account the momentary local conditions. This is also a reason for GSM not having standardized a uniform algorithm for the determination of the moment of the handover. For this decision about when to perform a handover, network operators can develop and deploy their own algorithms which are optimally tuned for their networks. This is made possible through standardizing only the signaling interface that defines the processing of the handover and through transferring the handover decision to the BSS. The GSM handover is thus *a network-originated handover* as opposed to a *mobile-originated handover*, where the handover decision is made by the mobile station. An advantage of this handover approach is that the software of the mobile station need not be changed when the handover strategy or the handover decision algorithm is changed in all or parts of the network. Even though the GSM standard does not prescribe a mandatory handover decision algorithm, a simple algorithm is proposed, which can be selected by the network operator or replaced by a more complex algorithm.

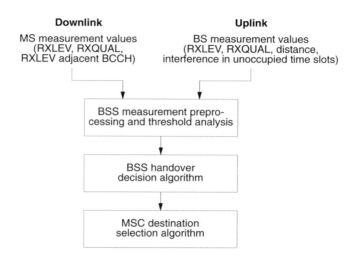

Figure 8.16: Decision steps in a GSM handover

In principle, a GSM handover always proceeds in three steps (Figure 8.16), which are based on the measurement data provided by the mobile station over the SACCH, and on the measurements performed by the BSS itself. Foremost among these data items are the current channel's received signal level (RXLEV) and the signal quality (RXQUAL), both on the uplink (measured by the BSS) and on the downlink (measured by the MS). In order to be able to identify neighboring cells as potential targets for a handover, the mobile station measures in addition the received signal level RXLEV_CELL(n) of up to 16 neighboring base stations. The RXLEV values of the six base stations which can be received best are reported every 480 ms to the BSS. Further criteria for the handover decision algorithm are the distance between MS and BTS measured via the *Timing Advance* (TA) of the *Adaptive Frame Alignment* (see Section 5.3.2) and measurements of the interference in unused time slots. A new value of each of these measurements is available every 480 ms.

Measurement preprocessing calculates average values from these measurements, whereby at least the last 32 values of RXLEV and RXQUAL must be averaged. The resulting mean values are continuously compared with thresholds (see Table 8.1) after every SACCH interval.

These threshold values can be configured individually for each BSS through management interfaces of the OMSS (see Section 3.3.4). The principle used for the comparison of measurements with the threshold is to conduct a so-called Bernoulli experiment: if out of the last N_i mean values of a criterion i more than P_i go under (RXLEV) or over (RXQUAL, MS_RANGE) the threshold, then a handover may be a necessary. The values of N_i and P_i can also be configured through network management. Their allowed range is defined as the interval [0; 31].

In addition to these mean values, a BSS can calculate the current power budget PBGT(**n**), which represents a measure for the respective path loss between mobile station and current base station or a neighboring base station n. Using this criterion, a handover can always be caused to occur to the base station with the least path loss for the signals from or to the mobile station. The PBGT takes into consideration not only the RXLEV_DL of the current downlink and the RXLEV_NCELL(n) of the neighboring BCCH but also the maximal transmitter power P (see Table 5.8) of a mobile station, the maximal power MS_TXPWR_MAX allowed to a mobile station in the current cell, and the maximal power MS_TXPWR_MAX(n) allowed to mobiles in the neighboring cells. In addition, the calculation uses the value PWR_C_D, which is the difference between maximal transmitter power on the downlink and current transmitter power of the BTS in the downlink, a measure for the available power control reserve.

Table 8.1: Threshold values for the GSM handover

Threshold value	Typ. value	Meaning
L_RXLEV_UL_H	−103 to −73 dBm	Upper handover threshold of received signal level in uplink
L_RXLEV_DL_H	−103 to −73 dBm	Upper handover threshold of received signal level in downlink
L_RXLEV_UL_IH	−85 to −40 dBm	Lower(!) received signal level threshold in uplink for internal handover
L_RXLEV_DL_IH	−85 to −40 dBm	Lower(!) received signal level threshold in downlink for internal handover
RXLEV_MIN(n)	approx. −85 dBm	Minimum required RXLEV of BCCH of cell n to perform a handover to this cell
L_RXQUAL_UL_H	–	Lower handover threshold of bit error ratio in uplink
L_RXQUAL_DL_H	–	Lower handover threshold of bit error ratio in downlink
MS_RANGE_MAX	2 to 35 km	Maximum distance between mobile and base station
HO_MARGIN(n)	0 to 24 dB	Hysteresis to avoid multiple handovers between two cells

Thus the power budget for a neighboring base station n is calculated as follows:

$$PBGT(n) = (\ Minimum(\ MS\_TXPWR\_MAX, P\ ) - RXLEV\_DL - PWR\_C\_D\ )$$
$$- (\ Minimum(\ MS\_TXPWR\_MAX(n), P\ ) - RXLEV\_NCELL(n)\ )$$

A handover to a neighboring base station can be requested, if the power budget is PBGT(n)>0 and greater than the threshold HO_MARGIN(n). The causes for handover which are possible using these criteria are summarized in Table 8.2. As can be seen, the signal criteria of the uplink and downlink as well as the distance from the base station and power budget can lead to a handover.

The BSS makes a handover decision by first determining the necessity of a handover using the threshold values of Table 8.1. In principle, one can distinguish three categories:

- Handover because of more favorable path loss conditions
- Mandatory intercell handover
- Mandatory intracell handover

Table 8.2: Handover causes

Handover Cause	Meaning
UL_RXLEV	Uplink received signal level too low
DL_RXLEV	Downlink received signal level too low
UL_RXQUAL	Uplink bit error ratio too high
DL_RXQUAL	Downlink bit error ratio too high
PWR_CTRL_FAIL	Power control range exceeded
DISTANCE	MS to BTS distance too high
PBGT(n)	Lower value of path loss to BTS n

Situations where a neighboring base station shows more favorable propagation conditions and therefore lower path loss, do not necessarily force a handover. Such potential handover situations to a neighboring cell are discovered through the PBGT(n) calculations. To make a handover necessary, the power budget of the neighboring cell must be greater than the threshold HO_MARGIN(n).

The recognition of a mandatory handover situation (Figure 8.17) within the framework of the *Radio Subsystem Link Control* (see also Section 5.5 and Figure 5.19) is based on the received signal level and signal quality in uplink and downlink as well as on the distance between MS and BTS. Going over or under the respective thresholds always necessitates a handover.

Here are the typical situations for a mandatory handover:

- The received signal level in the uplink or downlink (RXLEV_UL / RXLEV_DL) drops below the respective handover threshold value (L_RXLEV_UL_H/ L_RXLEV_DL_H) and the power control range has been exhausted, i.e. the MS and/or the BSS have reached their maximal transmitter power (see Section 5.5.2).
- The bit error ratio as a measure of signal quality in uplink and/or downlink (RXQUAL_UL / RXQUAL_DL) exceeds the respective handover threshold value (L_RXQUAL_UL_H / L_RXQUAL_DL_H), while at the same time the received signal level drops into the neighborhood of the threshold value.
- The maximum distance to the base station (MAX_MS_RANGE) has been reached.

A handover can also become mandatory, even if the handover thresholds are not exceeded, if the lower thresholds of the transmitter power control are exceeded (L_RXLEV_xx_P / L_RXQUAL_xx_P, see Table 5.9), even though the maximum transmitter power has been reached already. The cause of handover indicated is the failure of the transmitter power control (PWR_CTR_FAIL, see Table 8.2).

A special handover situation exists, if the bit error ratio RXQUAL as a measurement for signal quality in uplink and/or downlink exceeds its threshold and at the same time the received signal level is greater than the thresholds L_RXLEC_UL_IH / L_RXLEC_DL_IH. This strongly hints at an existing severe cochannel interference. This problem can be solved with an (internal) intracell handover, which the BSS can perform on its own without support from the MSC. It is also considered as a mandatory handover.

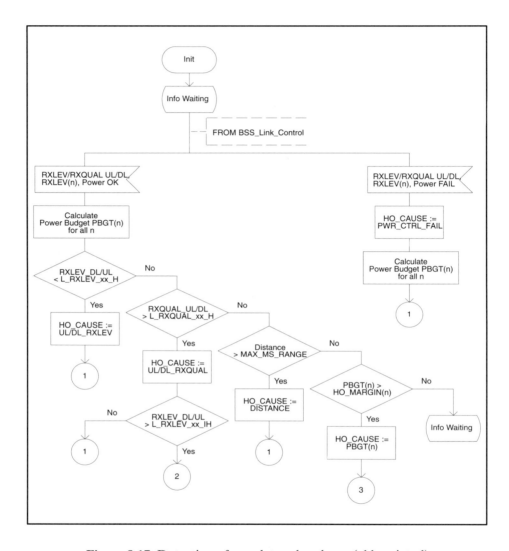

Figure 8.17: Detection of mandatory handover (abbreviated)

If the BSS has detected a handover situation, a list of candidates as possible handover targets is assembled using the BSS decision algorithm. For this purpose, one first determines which BCCH of the neighboring cell n is received with sufficient signal level:

$$\text{RXLEV\_NCELL}(n) > (\ \text{RXLEV\_MIN}(n)$$
$$+ \text{Maximum}(\ 0, (\text{MS\_TXPWR\_MAX}(n) - P\ )\ )\ )$$

The potential handover targets are then assembled in an ordered list of preferred cells according to their path loss compared to the current cell (Figure 8.18). For this purpose, the power budget of the neighboring cells in question is again evaluated:

$$PBGT(n) - \text{HO\_MARGIN}(n) > 0$$

All cells n which are potential targets for a handover due to RXLEV_NCELL(n) and lower path loss than the current channel are then reported to the MSC with the message HANDOVER REQUIRED (Figure 8.18) as possible handover targets. This list is sorted by priority according to the difference (PBGT(n) − HO_MARGIN(n)). The same message HANDOVER REQUIRED is also generated if the MSC has sent a message HANDOVER CANDIDATE ENQUIRY to the BSS.

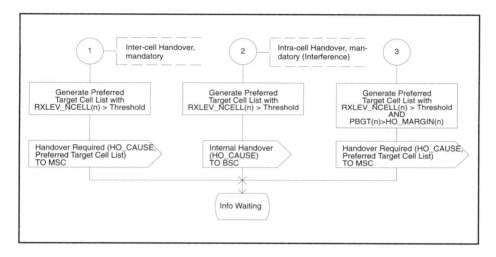

Figure 8.18: Completion of handover decision in the BSS

The conditions at a cell boundary in the case of exhausted transmitter power control (PWR_C_D = 0) are shown in Figure 8.19 with a mobile station moving from the current cell to a cell B. The threshold RXLEV_MIN(B) is reached very early; however, the handover is somewhat moved in the direction of cell B because of the positive HO_MARGIN(B) for the power budget. When moving in the opposite direction, the handover would be delayed in the other direction due to HO_MARGIN(A) of cell A. This has the effect of a hysteresis which reduces repeated handovers between both cells due to fading (ping-pong handover).

Besides varying radio conditions (fading due to multipath propagation, shadowing, etc.) there are many other sources of error with this kind of handover. Recognize, on one hand, that there are substantial delays between measurement and reaction due to the averaging process. This leads to executing the handover too late on a few occasions. It is more important, however, that the current channel is compared with the BCCH of the neighboring cells rather than the traffic channel to be used after the handover decision, which could suffer from different propagation conditions (frequency-selective fading etc.).

Finally, the MSC decides about the target cell of the handover. This decision takes into consideration the following criteria in decreasing order of priority: handover due to signal quality (RXQUAL), received signal level (RXLEV), distance, and path loss (PBGT). This prioritization is especially effective when there are not enough traffic channels available and handover requests are competing for the available channels.

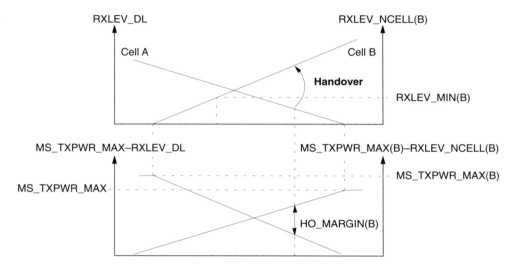

Figure 8.19: Handover criteria for exhausted transmitter power control

The standard explicitly points out that all measurement results must be sent with the message HANDOVER REQUIRED to the MSC, so that in the end the option remains open to implement the complete handover decision algorithm in the MSC.

8.4.4 MAP and Inter-MSC Handover

The most general form of handover is the inter-MSC handover. The mobile station moves over a cell boundary and enters the area of responsibility of a new MSC. The handover caused by this move requires communication between the involved MSCs. This occurs through the SS#7 using transactions of the *Mobile Application Part* (MAP).

8.4.4.1 Basic Handover between two MSCs

The principal sequence of operations for a basic handover between two MSCs is shown in Figure 8.20. The MS has indicated the conditions for the handover, and the BSS requests the handover from MSC-A (HANDOVER REQUIRED). MSC-A decides positively for a handover and sends a message PERFORM HANDOVER to MSC-B. This message contains the necessary data to enable MSC-B to reserve a radio channel for the MS. Above all, it identifies the BSS which is to receive the connection. MSC-B assigns a handover number and tries to allocate a channel for the MS. If a channel is available, the response RADIO CHANNEL ACKNOWLEDGE contains the new MSRN to the MS and the designation of the new channel. If no channel is available, this is also reported to MSC-A which then terminates the handover procedure.

When a RADIO CHANNEL ACKNOWLEDGE is successful, an ISDN channel is switched through between the two MSCs (ISUP messages IAM and ACM), and both MSCs send an acknowledgment to the MS (HA INDICATION, HB INDICATION). The MS then resumes the connection on the new channel after a short interruption (HB CONFIRM). MSC-B then sends a message SEND END SIGNAL to MSC-A and thus causes the release of the old radio connection. After the end of the connection (ISUP messages REL, RLC), MSC-A generates a message END SIGNAL for MSC-B which then sends a HANDOVER REPORT to its VLR.

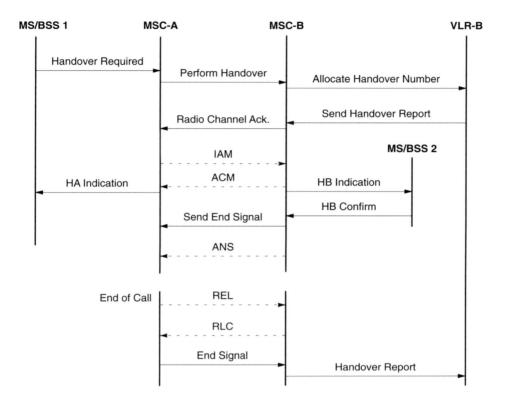

Figure 8.20: Principal operation of a basic handover

8.4.4.2 Subsequent Handover

After a first basic handover of a connection from MSC-A to MSC-B, a mobile station can move on freely. Further intra-MSC handovers can occur (Figure 8.15), which are processed by MSC-B.

If, however, the mobile station leaves the area of MSC-B during this connection, a *Subsequent Handover* becomes necessary. Two cases are distinguished: in the first case, the mobile station returns to the area of MSC-A, whereas in the second case it enters the area of a new MSC, now called MSC-B'. In both cases, the connection is newly routed from MSC-A. The connection between MSC-A and MSC-B is taken down after a successful subsequent handover.

A subsequent handover from MSC-B back to MSC-A is also called *handback* (Figure 8.21). In this case, MSC-A, as the controlling entity, does not need to assign a handover number and can search directly for a new radio channel for the mobile station. If a radio channel can be allocated in time, both MSCs start their handover procedures at the air interface (HA / HB INDICATION) and complete the handover. After completion, MSC-A terminates the connection to MSC-B. The message END SIGNAL terminates the MAP process in MSC-B and causes a HANDOVER REPORT to be sent to the VLR of MSC-B; the ISUP message RELEASE releases the ISDN connection.

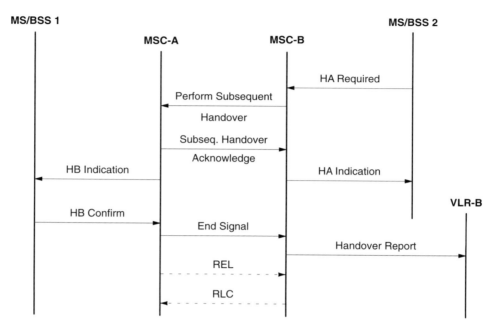

Figure 8.21: Principle of subsequent handover from MSC-B to MSC-A (handback)

The procedure for a subsequent handover from MSC-B to MSC-B′ is more complicated. This procedure consists of two parts:

- A *Subsequent Handover* from MSC-B to MSC-A
- A *Basic Handover* between MSC-A and MSC-B′

The principal operation of this handover is illustrated in Figure 8.22.

In this case, MSC-A recognizes from the message PERFORM SUBSEQUENT HANDOVER, sent by MSC-B, that it is a case of handover to an MSC-B′, and it initiates a *Basic Handover* to MSC-B′. MSC-A informs MSC-B after receiving the ISUP message ACM from MSC-B′ about the start of handover at MSC-B′ and thereby frees the handover procedure at the radio interface from MSC-B. Once MSC-A receives the MAP message SEND END SIGNAL from MSC-B, it considers the handover as complete, sends the message END SIGNAL to MSC-B to terminate the MAP procedure and cancels the ISDN connection.

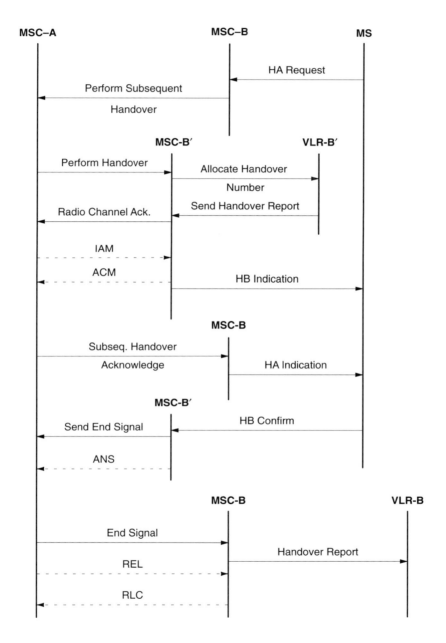

Figure 8.22: Principle of subsequent handover from MSC-B to MSC-B′

9 Data Communication and Networking

9.1 Reference Configuration

GSM was conceived in accordance with the guidelines of ISDN. Therefore, a reference configuration is also defined for GSM systems, similar to the one used in ISDN systems. Using the reference configuration, one gets an impression of the range of services and the kinds of interfaces to be provided by mobile stations. Furthermore, the reference configuration indicates at which interface which protocols or functions terminate and where adaptation functions may have to be provided.

The GSM reference configuration comprises the functional blocks of a mobile station (Figure 9.1) at the user–network interface Um. The mobile equipment is subdivided into a *Mobile Termination* (MT) and various combinations of *Terminal Adapter* (TA) and *Terminal Equipment* (TE), depending on the kind of service access and interfaces offered to the subscriber.

At the interface to the mobile network, the air interface Um, *Mobile Termination* (MT) units are defined. An integrated mobile speech or data terminal is represented only by an MT0. The MT1 unit goes one step further and offers an interface for standard-conforming equipment at the ISDN S reference point, which can be connected directly as end equipment. Likewise, normal data terminal equipment with a standard interface (e.g. V.24) can be connected via a TA and this way use the mobile transmission services. Finally, the TA functionality has been integrated into units of type MT2.

At the S or R reference point, the GSM bearer or data services are available (access points 1 and 2 in Figure 9.1), whereas the teleservices are offered at the user interfaces of the TE (access point 3, Figure 9.1). Among the bearer services besides the transmission of digitized speech, there are circuit-switched and packet-switched data transmission. Typical teleservices besides telephony are, for example, *Short Message Service* (SMS), Group 3 fax service, or emergency calls from anywhere.

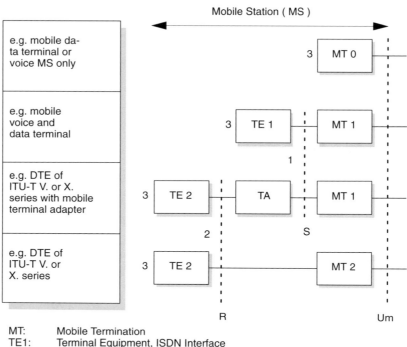

MT: Mobile Termination
TE1:   Terminal Equipment, ISDN Interface
TE2:   Terminal Equipment, no ISDN interface (V., X.)
TA: Terminal Adapter
Um: Reference Point Radio Interface
R/S: Reference Points ISDN / non-ISDN

Figure 9.1: GSM reference configuration

9.2   Overview of Data Communication

Besides voice communication, GSM networks offer a series of data services and teleservices which are based on them. Voice service needs only a switched-through physical connection, which changes its bit rate in the BSS due to the speech transcoding in the TRAU. From the MSC on, the speech signals in GSM networks are transported in standard ISDN format with a bit rate of 64 kbit/s. In comparison, realizing data services and the other teleservices like Group 3 fax is considerably more complicated. Because of the psychoacoustic compression procedures of the GSM speech codec, data cannot be simply transmitted as a voiceband signal as in the analog network — a complete reconstruction of the data signal would not be possible. Therefore, a solution to digitize the voiceband signal similar to ISDN is not possible. Rather the available digital data must be transmitted in unchanged digital form by avoiding speech codecs in the PLMN, as is possible in the ISDN. Here we have to distinguish two areas where special measures have to be taken: first, the realization of data and teleservices at the air interface or within the mobile network, and second, at the transition between mobile and fixed network with the associated mapping of service features. These two areas are illustrated schematically in Figure 9.2.

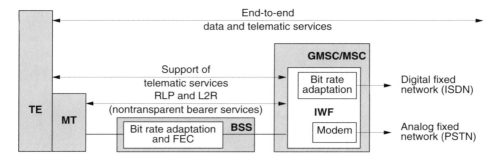

Figure 9.2: Bearer services, interworking, and teleservices

A PLMN offers transparent and nontransparent services. These bearer services carry data between the *Mobile Termination* (MT) of the mobile station and the *Interworking Function* (IWF) of the MSC. For the realization of bearer services, the individual units of the GSM network define several functions:

- *Bit Rate Adaptation* (RA)
- *Forward Error Correction* (FEC)
- ARQ *Error Correction* with the *Radio Link Protocol* (RLP)
- Adaptation protocol *Layer 2 Relay* (L2R)

For the transmission of transparent and nontransparent data, several rate adaptation stages are required to adapt the bit rates of the bearer services to the channel data rates of the radio interface (traffic channels with 3.6 kbit/s, 6 kbit/s, and 12 kbit/s) and to the transmission rate of the fixed connections. A bearer service for data transmission can be realized in the following two ways: 9.6 kbit/s data service requires a full-rate traffic channel, all other data services can either be realized on a full-rate or half-rate channel. A mobile station must support both types of data traffic channels, independent of what is used for speech transmission. The data signals are transcoded first from the user data rate (9.6 kbit/s, 4.8 kbit/s, 2.4 kbit/s, etc.) to the channel data rate of the traffic channel, then further to the data rate of the fixed connection between BSS and MSC (64 kbit/s) and finally back to the user data rate. This bit rate adaptation (RA) in GSM corresponds in essence to the bit rate adaptation in the ITU-T standard V.110, which specifies the support of data terminals with an interface according to the V. series on an ISDN network [17].

On the radio channel, data is protected through the forward error correction procedures (FEC) of the GSM PLMN; and for nontransparent data services, data is additionally protected by the ARQ procedure of RLP on the whole network path between MT and MSC. Thus RLP is terminated in the MT and MSC. The protocol adaptation to RLP of Layers 1 and 2 at the user interface is done by the *Layer 2 Relay* (L2R) protocol.

Finally, the data is passed on from MSC or GMSC over an *Interworking Function* (IWF) to the respective data connection. The bearer services of the PLMN are transformed to the bearer services of the ISDN or another PLMN in the IWF, which is usually activated in an MSC near the MS, but could also reside in the GMSC of the network transition. In the case of ISDN this transition is relatively simple, since it may just require a potential bit rate adaptation. In the case of an analog PSTN, the available digital data must be transformed by a modem into a voiceband signal, which can then be transmitted on an analog voiceband of 3.1 kHz.

The bearer services realized in this way can offer the protocols that may be required for the support of teleservices between TE and IWF. An example is the fax adaptation protocol. The fax adapter is a special TE which maps the Group 3 fax protocols with their analog physical interface upon the digital bearer services of a GSM PLMN. Thus, after another adaptation into an analog fax signal in the IWF of the MSC, it enables the end-to-end transfer of fax messages according to the ITU-T Standard T.30.

A possible interworking scenario for transparent data services of GSM with transition to a PSTN is shown in Figure 9.3. The analog circuit-switched connection of the PSTN represents a transparent channel which can be used to transport arbitrary digital data signals in the voiceband. In the analog network, a subscriber selects telephone or modem depending on whether he or she wants to transmit speech or data. In the PLMN, however, the channel coding has to be changed for different services (error protection for different bearer services, see Section 6.2). The bit rate adaptation has to be activated and the speech coding deactivated. In the IWF of the MSC, besides the bit rate adaptation, a modem needs to be added for data communication with the partner in the fixed network. In the GSM network, voice signals therefore take a different path than data signals; in the case shown in Figure 9.3, the data signals are directed from the IWF to the modem, where they are digitized, then passed on after bit rate adaptation to transmission on the radio channel. In the opposite direction, the IWF passes the PCM-coded information on an ISDN channel (64 kbit/s) to the GMSC. From there it is transformed into an analog signal in a network transition switching unit and carried as a voiceband signal in the PSTN to the analog terminal.

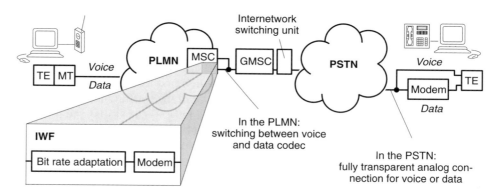

Figure 9.3: Interworking scenario PLMN-PSTN for transparent data services

After these introductory remarks, the GSM data and teleservices and their realization will be discussed in more detail in the following sections.

9.3   Service Selection at Transitions between Networks

A specific interworking problem arises for data services between PLMN and ISDN / PSTN networks. Mobile-terminated calls require that the calling subscriber (ISDN or PSTN subscriber) tells the GMSC which service (speech, data, fax, etc.) he or she wants to use. In

ISDN, a *Bearer Capability* (BC) information element would have to be included in the SETUP message. This BC information element could then be passed on by the network transition switching unit to the GMSC and from there to the local MSC, which could thus activate the required resources. In the course of call processing (CC, see Section 7.4.5), the mobile station would also be informed about the kind of service requested by the calling subscriber and could activate the needed functions. The calling subscriber, however, if there is no ISDN signaling as in analog networks, is not able to do this kind of BC signaling. The service selection therefore has to use another mechanism. The GSM standard proposes two possible solutions, which are always to be used for service selection independent of the type of originating network (ISDN or PSTN).

- *Multinumbering*
  The home network with this option assigns to each mobile subscriber several MSISDN numbers, each with a specific *Bearer Capability* (BC), which can be obtained at each call from the HLR. This way the service that an incoming call wants is always uniquely determined. The BC information element is given to the mobile station when the call is being set up, so the MS can decide based on its technical features whether it wants to accept the call.

- *Single numbering*
  Only a single MSISDN is assigned to the mobile subscriber, and there is no BC information element transmitted with an incoming call. The MS recognizes then that a specific BC is needed when a call is accepted and requests the BC from the MSC. If the network is able to offer the requested service, the call is switched through.

Usually, the multinumbering solution is favored, since one can already verify at call arrival time in the MSC whether the requested resources are available, and the MSC side can decide about accepting the call. There is no negotiation about the BC between MS and MSC, so no radio resources are occupied unnecessarily, and the call set-up phase is not extended.

9.4   Bit Rate Adaptation

Five basic traffic channels are available in GSM for the realization of bearer services: TCH/H2.4, TCH/H4.8, TCH/F2.4, TCH/F4.8, TCH/F9.6 (see Table 5.2 and Table 6.2) with bit rates of 3.6 kbit/s, 6 kbit/s, and 12 kbit/s. The bearer services (Table 4.2) with bit rates from 300 bit/s up to 9.6 kbit/s must be realized on these traffic channels. Furthermore, on the fixed connections of the GSM network, the data signals are transmitted with a data rate of 64 kbit/s.

The terminals connected at reference point R have the conventional asynchronous and synchronous interfaces. The data services at these interfaces work at bit rates as realized by GSM bearer services. Therefore, the data terminals at the R reference point have to be bit rate adapted to the radio interface. This bit rate adaptation is derived from the V.110 standard used in ISDN in which the bit rates of the synchronous data streams are going through a two-step procedure; first, frames are formed at an intermediate rate which is a multiple of 8 kbit/s; this stream is converted to the channel bit rate of 64 kbit/s [22]. The asynchronous services are preprocessed by a stuffing procedure using stop bits to form a synchronous data stream.

A V.110 procedure modified according to the requirements of the air interface is also used in GSM. In essence, GSM performs a transformation of the data signals from the user data rate (e.g. 2.4 kbit/s or 9.6 kbit/s) at the R reference point to the intermediate data rate (8 kbit/s or 16 kbit/s) and finally to the ISDN bit rate of 64 kbit/s. The adaptation function from user to intermediate rate is called RA1; the adaptation function from intermediate rate to ISDN is called RA2. A GSM-specific bit rate adaptation step is added between the intermediate rate and the channel data rate (3.6 kbit/s, 6 kbit/s, or 12 kbit/s) of the traffic channel at the reference point Um of the air interface. This adaptation function from intermediate to channel bit rate is designated as RA1/RA1′. An adaptation function RA1′ performs the direct adaptation from user to channel data rate without going through the intermediate data rate. Table 9.1 gives an overview of the bit rates at the reference points and the intermediate data rates between the RA modules.

Table 9.1: Data rates for GSM bit rate adaptation

Interface	Data rate		Interface	
	User	Intermediate	Radio	S
Reference point	R	–	Um	S
RA1	≤ 2.4 kbit/s	8 kbit/s		
RA1	4.8 kbit/s	8 kbit/s		
RA1	9.6 kbit/s	16 kbit/s		
RA2		8 kbit/s		64 kbit/s
RA2		16 kbit/s		64 kbit/s
RA1/RA1′		8 kbit/s	3.6 kbit/s	
RA1/RA1′		8 kbit/s	6 kbit/s	
RA1/RA1′		16 kbit/s	12 kbit/s	

Adaptation frames are defined for the individual bit rate adaptation steps. These frames contain signaling and synchronization data besides the user data. They are defined based on V.110 frames, and one distinguishes three types of GSM adaptation frames according to their length (36 bits, 60 bits, and 80 bits) as shown in Figure 9.4 and Figure 9.5.

The conversion of data signals from user to intermediate rate in the RA1 stage uses the regular 80-bit frame of the V.110 standard. In this adaptation step, groups of 48 user data bits are supplemented with 17 fill bits and 15 signaling bits to form an 80-bit V.100 frame. Because of the ratio 0.6 of user data to total frame length, this adaptation step converts user data rates of 4.8 kbit/s into 8 kbit/s and from 9.6 kbit/s to 16 kbit/s. All user data frames of less then 4.8 kbit/s are "inflated" to a data signal of 4.8 kbit/s by repeating the individual data bits; for example, a 2.4 kbit/s signal all bits are doubled, or with a 600 bit/s signal the bits are written eight times into an RA1 frame.

At the conversion of the intermediate data rate to the channel data rate in the RA1/RA1′ stage, the 17 fill bits and 3 of the signaling bits are removed from the RA1 frame, since they are only used for synchronization and not needed for transmission across the air interface. This yields a modified V.100 frame of length 60 bits (Figure 9.5), and the data rate is adapted from 16 kbit/s to 12 kbit/s or from 8 kbit/s to 6 kbit/s, respectively.

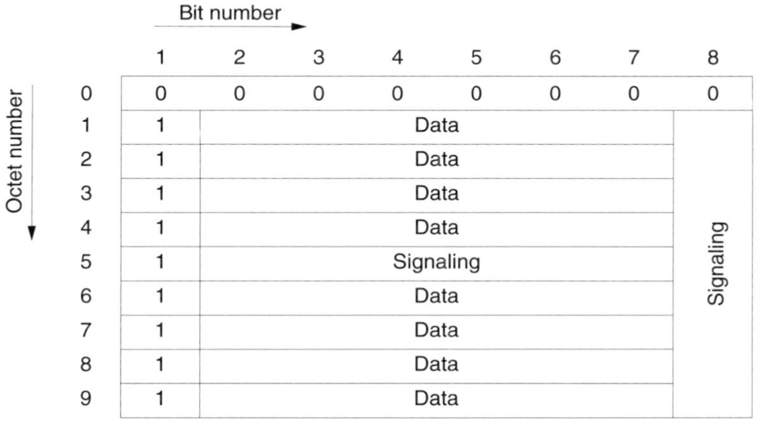

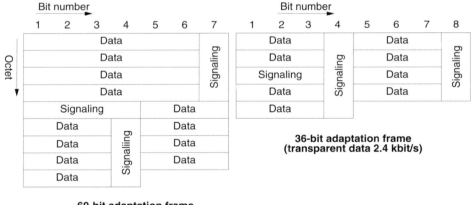

Figure 9.4: V.110 80-bit adaptation frame for the RA1 stage

Figure 9.5: Modified V.110 adaptation frame for the RA1′ stage

In the case of user data rates of 4.8 kbit/s or 9.6 kbit/s, adaptation to the channel data rate is already complete. Only for user data rates of less than 4.8 kbit/s do additional parts of the multiple user bits need to be removed, which results in a modified V.110 frame of 36 bits. Thus the user data rates of less than 4.8 kbit/s are adapted to a channel data rate of 3.6 kbit/s. The user data bits of a 2.4 kbit/s signal are then not transmitted twice anymore, or the 600 bit/s user data signals are only written four times into the frames of the RA1′ stage. This is, however, only true for the transparent bearer services. For the non-transparent bearer services, the modified 60-bit V.110 frame is used completely for the transmission of the 60 data bits of an RLP PDU. The required signaling bits are multiplexed with user data into the RLP frame through the Layer 2 Relay protocol L2R.

The modem used for communication over the PSTN resides in the IWF of the MSC, since data is transmitted from here on in digital form within the PLMN. For congestion and flow control and other functions at the modem interface, the interface signals must therefore be carried from the modem through the PLMN to the mobile station. For this purpose, signaling bits are reserved in the frames of the bit rate adaptation function, which represent these signals and thus give the MS direct modem control. The connection of such a bearer service is therefore transparent not only for user data, but also for out-of-band signaling of the (serial) modem interface in the IWF.

9.5   Asynchronous Data Services

Asynchronous data transmission based on the V. and X. series interfaces is widespread in fixed networks. In order to support such "non-GSM" interfaces, the mobile station can include a *Terminal Adapter* (TA) over which standard terminals with a V. or X. interface (e.g. V.24) can be connected. Such an adaptation unit can also be integrated into the mobile station (MT2, Figure 9.1).

Flow control between TA and IWF can be supported in different ways, just as in ISDN:

- *No flow control*; it is handled end-to-end in higher protocol layers (e.g. transport layer)
- *Inband Flow Control* with X-ON / X-OFF protocol
- *Out-of-Band Flow Control* according to V.110 through interface leads 105 and 106

9.5.1   Transparent Transmission in the Mobile Network

In the case of transparent transmission, data is transmitted with pure Layer 1 functionality. Besides error protection at the air interface, only bit rate adaptations are performed.

User data is adapted to the traffic channel at the air interface according to the data rate and protected with forward error-correcting codes (FECs) against transmission errors. As an example, Figure 9.6 shows the protocol model for transparent asynchronous data transmission over an MT1 with an S interface. Data is first converted in the TE1 or TA into a synchronous data stream by bit rate adaptation (stage RA0). In further stages, data rates are adapted with an MT1 to the standard ISDN (RA1, RA2), and then converted in MT1 over RA2, RA1, and RA1' to the channel bit rate at the air interface. Provided with an FEC, the data is transmitted and then converted again in the BSS by the inverse operations of bit rate adaptation to 64 kbit/s at the MSC interface. But much more frequently than an MT1 with an (internal) S interface, mobile stations realize a pure R interface without internal conversion to the full ISDN rate in the RA2 stage. This avoids the bit rate adaptation step RA2 and thus the conversion to the intermediate data rate in the RA1 stage. The signal is converted immediately after the asynchronous–synchronous conversion in the RA0 stage from the user data rate to the channel data rate (stage RA1').

A variation without terminal adapter is shown schematically in Figure 9.7. Here the complete interface functionality, *Interface Circuit* (I/Fcct), for a serial V. interface is integrated with the required adaptation units. The data signals D are converted into a synchronous signal in MT2 (RA0), packed into a modified V.100 frame together with signaling information S from the V.-interface, and adapted to the channel data rate (RA1′). After FEC, the data signals are transmitted over the air interface and finally converted for further transmission to the data rate of an ISDN B channel after decoding and potential error correction in the BSS (RA2).

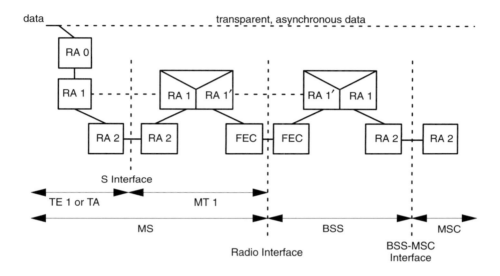

Figure 9.6: Transparent transmission of asynchronous data in GSM

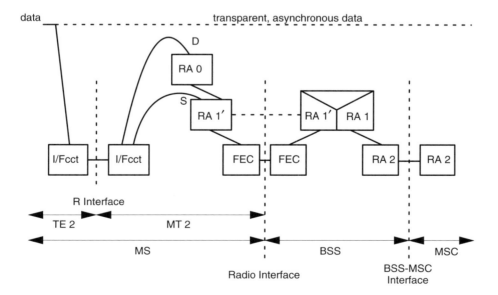

Figure 9.7: Transparent transmission of asynchronous data across the R interface

Figure 9.8 shows a complete scenario with all appropriate network transitions for a transparent bearer service with modem in the interworking function for the conversion of the digital data signals into an analog voiceband signal. A mobile data terminal uses the transparent bearer service of a GSM PLMN over an R interface (S interface is also possible). The data is circuit-switched to the IWF in the MSC. To communicate with a modem in the fixed network, the IWF activates an appropriate modem function and converts the digital data signals into an analog voiceband signal. The IWF digitizes this voiceband signal again and passes the data on in PCM-coded format through the GMSC. After the network transition, the data signal is finally transmitted to the modem of the communication partner. This modem can be within a terminal in the PSTN or belong to an ISDN terminal. Before being transmitted in the PSTN, the PCM-coded signal is again converted into an analog voiceband signal. In ISDN the signal is transmitted as a PCM-coded signal of category *3.1 kHz Audio*; a repeated conversion is not necessary. An ISDN subscriber needs an adaptation unit TA' for the conversion of the digital voiceband signal into an analog signal, which can then be processed further with a modem and passed on to the data terminal.

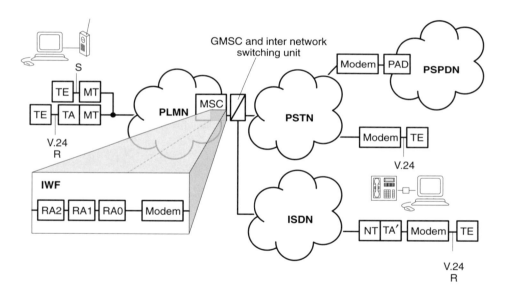

Figure 9.8: Principle of transparent asynchronous data transfer (variant with modem)

Another variant consists of a circuit-switched modem connection to a packet-switched network access node, as is possible from fixed connection ports. In these access nodes, the asynchronous modem signals are combined into packets in a *Packet-Assembler / Disassembler* (PAD) module and then transmitted through the packet-switched network. This variant of packet network access has the disadvantage that one has to switch through to the PAD over a long path, especially in the case of international roaming, since usually the nearest PAD is not the one allowed to the subscriber for access to packet networks.

It is also possible to connect to standard ISDN terminals without an analog modem based on the digital data transmission capability of ISDN. For this purpose, the transmission mode *Unrestricted Digital* has been defined. In this case, there is only a bit rate adaptation according to V.110 (Figure 9.9). The data arrives at the MSC from the BSS in V.110 frames on an ISDN channel with 64 kbit/s and transparently is passed on to the ISDN using a B channel again in V.110 frames. The otherwise necessary modems are entirely unnecessary in the case of *Unrestricted Digital* connections. However, an ISDN subscriber can connect a terminal through an analog modem by using an adaptation unit TA′ which converts the unrestricted digital signal into a voiceband signal according to one of the V. standards.

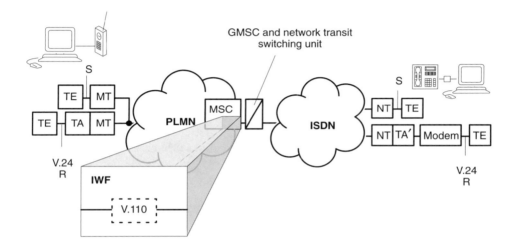

Figure 9.9: Transparent data transfer to an ISDN (unrestricted digital)

The quality of transparent data services in GSM varies with the radio field conditions. This is illustrated by examples of comparative field measurements of the transparent data service BS26 with 9.6 kbit/s data rate, comparing a moving and a standing mobile station. Figure 9.10 shows the weighted distribution of the bit errors of these two cases for a block length of 1024 bits. The weighted distribution indicates the frequency with which m bit errors occur in a block of length n bits (here $n=1024$):

$$P(m, n) = \sum_{i=m}^{\infty} P(\, i \text{ errors in } n\text{-bit-block}\,)$$

The distribution shown in Figure 9.10 represents measurements of the error statistics of BS26 which were performed in 1994 in a suburban area for moving and standing mobile stations [4]. Notice that the error frequency for the standing mobile station (stationaer) is clearly lower than for the moving mobile station (mobil). This result is obtained by averaging measurements over several locations and measurement tours. The resulting mean shows, for the moving station, a sometimes heavily varying channel due to the unavoidable fading phenomena, which frequently cause bursts with high bit error ratios, again resulting in an aggregate higher mean bit error ratio and thus also a higher packet error ratio $P(1, 1024)$.

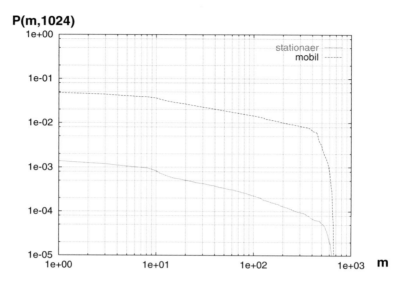

Figure 9.10: Weighted distribution of a transparent GSM bearer service (BS26)

9.5.2 Nontransparent Data Transmission

In contrast to the transparent transmission mode, the nontransparent mode in GSM data services protects the user data within the PLMN through a Layer 2 protocol, the *Radio Link Protocol* (RLP), in addition to the FEC procedures (convolutional coding, interleaving). This protection protocol further reduces the residual bit error ratio of data transmission in the mobile network. However, the automatic repeat requests (ARQ) of the RLP introduces additional transmission delays on the data path, and the effective user data throughput is reduced (protocol overhead).

User data between MT and MSC / IWF is protected on Layer 2 by the RLP. Two kinds of transmission errors are corrected this way: first, those caused by radio interference and remaining uncorrected by FEC, and second, those caused through the interruptions of handover. For signaling on the FACCH, time slots are "stolen" from the data traffic channel, which can cause data losses. The RLP protects user data against such losses, too.

For nontransparent data transfer with RLP, an additional sublayer in Layer 2 is required, the *Layer 2 Relay* (L2R) protocol. This relay protocol maps user data and status information of the IWF user–modem interface onto the information frames of the RLP. Depending on the kind of user data (character- or bit-oriented), one of two variations of the L2R is used: *Layer 2 Relay Bit Oriented Protocol* (L2RBOP), or *Layer 2 Relay Character Oriented Protocol* (L2RCOP). An L2R PDU is handed to RLP as a service data unit (SDU) and inserted into the RLP frame as a data field. The first octet of an L2R PDU always contains control information, like the status of the signaling lines of the serial interface. Beyond that, the L2R PDU can contain an arbitrary number of such status octets. They are always inserted into the user data stream when the state of the interface changes (e.g. hardware flow control). Thus, of the 200 user data bits (Figure 7.10), only a maximum of 192 can be used for payload. However, since the signaling information is already contained in the L2R PDU, these bits must not be considered for bit rate adaptation. Thus the modified 60-bit V.110 frame (Figure 9.5) can be completely occupied with data bits, and the full channel rate of maximum 12 kbit/s can be used for the transmission of RLP data. Next come a set of four modified V.110 frames which carry a complete RLP frame. Considering the protocol overhead of RLP (16.7%) and the minimum overhead of an L2R PDU (0.5%), one obtains a usable subscriber data rate of up to 9.95 kbit/s.

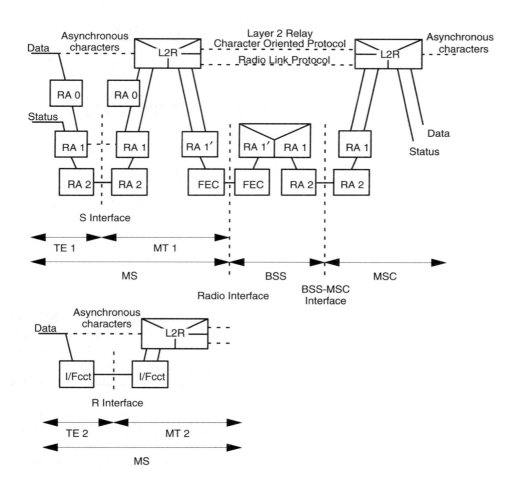

Figure 9.11: Nontransparent data transmission in GSM

The protocol model for asynchronous nontransparent character-oriented data transmission over the S interface in GSM will now be presented as an example (Figure 9.11). Data is transmitted by L2R protocol and *L2R Character Oriented Protocol* (L2RCOP) and RLP from the MT1 termination to the MSC. In between, as in the transparent case, are FEC, RA1, RA1′, and RA2. The RLP frames are transported in a synchronous mode. Only the user data stream at the user interface is asynchronous, and in the case of the model in Figure 9.11 the user data stream must be converted for the S interface into a synchronous data stream (RA0). In the case of a terminal with V. interface and an MT2 termination (reference point R), this bit rate adaptation at the S interface would be avoided. The asynchronous data is then directly accepted at the serial interface by the L2R (I/Fcct); potential start/stop bits are removed, and data are combined into L2R PDUs.

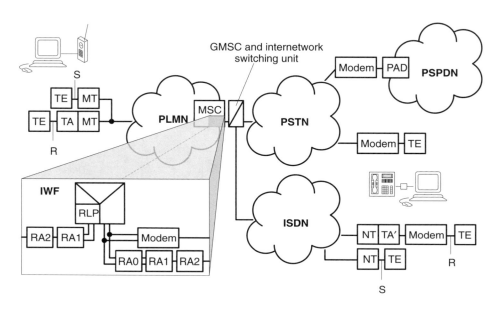

Figure 9.12: Principle of nontransparent data transfer

A complete scenario for network transition with nontransparent asynchronous GSM data services is shown in Figure 9.12. The RLP is terminated in the MSC / IWF, and user data are converted again into an asynchronous data stream by the associated L2R. For the nontransparent case too, the IWF offers both variants for the network transition: first, using modems and, second, *Unrestricted Digital*. In the case of a network transition to the PSTN or to a *3.1 kHz audio* connection into ISDN, the respective modem function is inserted and passes PCM-coded data to the GMSC (not shown in Figure 9.12), which directs them into the fixed networks. Using several bit rate adaptation steps (RA0 − RA1 − RA2, Figure 9.12), the user data can also be converted in the IWF into a synchronous *Unrestricted Digital* signal, which is then carried transparently over an ISDN B channel with 64 kbit/s.

9.5.3 PAD Access to Public Packet-Switched Data Networks

9.5.3.1 Asynchronous Connection to PSPDN PADs

As shown in Figures 9.8 and 9.12, access to *Packet Switched Public Data Networks* (PSPDNs), e.g. Accunet in the USA or Datex-P in Germany, is already possible using the asynchronous services of GSM. This requires a *Packet Assembler / Disassembler* (PAD) in the PSPDN, which packages the asynchronous data on the modem path into X.25 packets and also performs the reverse operation of unpacking. PAD access uses the protocols X.3, X.28 and X.29 (Triple X Profile). Just as from the fixed network, the mobile subscriber dials the extension of a PAD for access to the service of the PSPDN, provided packet network access is allowed. In this way, the subscriber has the same kind of access to the packet network as a subscriber from the fixed network, aside from the longer transmission delays and the higher bit error ratios. It is therefore recommended to transmit data for PAD access across the air interface in the PLMN in nontransparent mode with RLP [7].

9.5.3.2   Dedicated PAD Access in GSM

However, direct access to packet data networks through the asynchronous GSM data services has disadvantages:

- One needs another subscription to a packet data network operator besides to GSM.
- Independent of the current mobile subscriber's location, a circuit-switched connection to a PAD of a packet service provider is needed. Sometimes the packet network access is only allowed to specific PADs. This is a particular disadvantage if the mobile subscriber is currently in a foreign GSM network and incurs fees for international lines.

Therefore, GSM has defined another PSPDN access without these disadvantages: *Dedicated PAD Access* (Figure 9.13). The services are defined as *Bearer Services* BS41 through BS46 (Table 4.2). With this kind of PSPDN access from a PLMN, each PLMN has at least one PAD that is responsible for the packaging / unpackaging of the X.25 packets of the respective mobile subscriber.

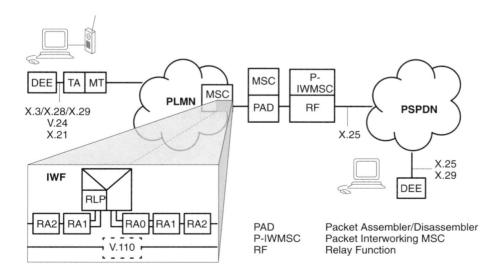

Figure 9.13: Dedicated PAD access through asynchronous GSM data services

For this purpose, a PAD is activated as an additional resource in the IWF or in an especially reserved MSC (Figure 9.13). This PAD can be reached in asynchronous mode again over transparent or nontransparent PLMN connections. However, with this solution the connection to the PAD is as short as possible, since a PAD is already reached in the nearest IWF, and international lines are never occupied for PAD access. Packetization of user data is already performed within the mobile network rather than in the remote PAD of a packet network operator, hence there is no need for a separate subscriber agreement with this packet network operator. The dedicated PAD of the current PLMN is now responsible for the packaging / unpackaging of the asynchronous data into / from X.25 packets, which are then passed on through a specific interworking MSC (P-IWMSC) to a public PSPDN (e.g. Transpac in France or Sprint's Telenet in the USA) [7].

The dedicated PAD has a uniform profile in all GSM networks; it is reached in each network with the same access procedure. Even in foreign PLMNs, a mobile station therefore gets the earliest and lowest cost access to the packet data network. Charging occurs to the account of the GSM extension (MSISDN) of the mobile subscriber; a separate *Network User Identification* (NUI) for the PSPDN is not necessary. However, only outgoing packet connections are possible.

9.6   Synchronous Data Services

9.6.1   Overview

Synchronous data services allow access to synchronous modems in the PSTN or ISDN as well as to circuit-switched data networks. Such access is not very significant; however, synchronous data services are defined in GSM. The essential differences to the asynchronous data transmission procedures are in bit rate adaptation and modems. For a synchronous data service, no RA0 bit rate adaptation is needed (conversion from asynchronous to synchronous), since data is already in synchronous format. Instead, special synchronous modems are required in the IWF. Synchronous data services can only be offered in transparent mode, with the exception of access to X.25 packet networks, which are a significant application of synchronous data service in GSM.

9.6.2   Synchronous X.25 Packet Data Network Access

The protocol model shown in Figure 9.14 is the model for synchronous data transmission in nontransparent mode with the packet data access protocol according to the ITU-T standard X.25. Because of the nontransparent transmission procedure, the X.25 *Link Access Procedure B* (LAPB) must be terminated in the MT as well as in the IWF. Since LAPB of the X.25 protocol stack operates in a bit-oriented mode, the *Layer 2 Relay Bit Oriented Protocol* (L2RBOP) is required.

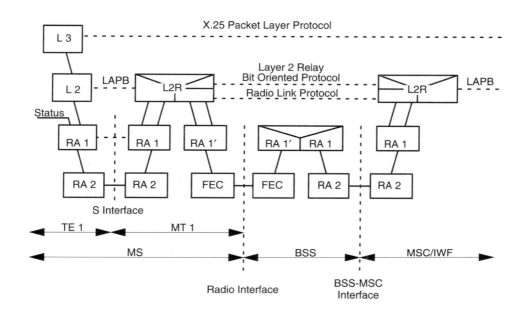

Figure 9.14: X.25 access at the ISDN S interface

9.6.2.1   Basic Packet Mode

Two variants of PSPDN access can be realized with the protocol model in Figure 9.14:

- PSPDN access according to ITU-T X.32
- Access to PSPDN packet handlers according to ITU-T X.31 Case A (basic packet mode).

PSPDN access according to X.32 has not met much acceptance with the applications; the X.31 procedure is used more often.

PSPDN access according to X.32 is the simpler variant. The X.25 packets can be transferred directly over a synchronous modem in the IWF to the PSPDN. This does not necessarily require nontransparent transmission in GSM, but it helps because of the lower bit error ratio. In case of the nontransparent transmission, the LAPB protocol has to be terminated in MT and IWF (see above). The subscriber needs a *Network User Identification* (NUI). Incoming and outgoing packet connections are possible, but again there is the problem of needing circuit-switched connections to the home PLMN, just like in case of international roaming.

The access procedure according to ITU-T standard X.31 Case A (basic packet mode) is the more favored variant of this group of services. The X.25 packets of the mobile subscriber are passed from the IWF to the packet handler of the ISDN. Since speed adaptation with the X.31 procedure is performed in the ISDN B channel by flag stuffing, the protocol has to be terminated in the IWF, which means that the nontransparent mode of GSM can be used. In this case too, there are connections possible only through the packet handler of the home network.

9.6.2.2 Dedicated Packet Mode

As in the case of asynchronous PAD access to packet data networks (see Section 9.5.3), the synchronous case of the X.25 access protocol also offers an alternative, which allows the most immediate transition to the PSPDN, even if the mobile station is in a foreign network (international roaming). For this purpose, a dedicated mode is also defined with each PLMN having its own packet handler.

Figure 9.15 shows the principle of this *Dedicated Packet Mode*. It essentially includes the functions of the basic packet mode, with the difference that the packet handler is integrated into the GSM network. Data is transmitted in the PLMN in nontransparent and synchronous mode. Access to the packet handler is the same in all PLMNs, and is also available to foreign mobile subscribers. Since packet data networks do not know roaming, only outgoing data calls are possible.

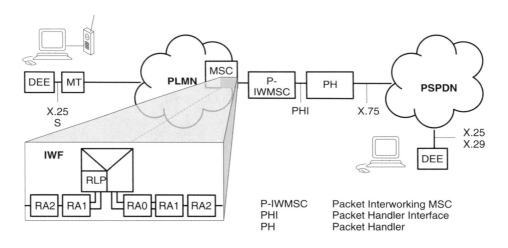

Figure 9.15: Dedicated packet mode with packet handler in GSM

9.7 Teleservices: Fax

Besides voice communication, the most important teleservices turned out to be *Short Message Service* and fax transmission. Realization of the fax service will be explained briefly in the following.

The GSM standard considers the connection of a regular Group 3 fax terminal with its two-wire interface to an appropriately equipped mobile station as the standard configuration of a mobile fax application. The GSM fax service is supposed to enable this configuration to conduct fax transmissions with standard Group 3 fax terminals over mobile connections. This requires mapping the fax protocol of the analog two-wire interface onto the digital GSM transmission, because the fax procedure defines a complete protocol stack with its own modulation, coding, user data compression, inband signaling, etc. Therefore the *Fax Adapter* (FA) has been defined for this mapping. The principle is summarized in Figure 9.16. The fax adapter of the mobile station converts the fax protocol of a standard Group 3 fax terminal on an analog two-wire line into a GSM internal fax adapter protocol. The PDUs of the adapter protocol are transmitted over the MT with the GSM data services to the fax adapter in the IWF of the MSC, and there they are again converted into the T.30 protocol on the analog line, or transmitted to the ISDN in PCM-coded format, where again a terminal adapter allows connection of a Group 3 fax terminal (not shown in Figure 9.16).

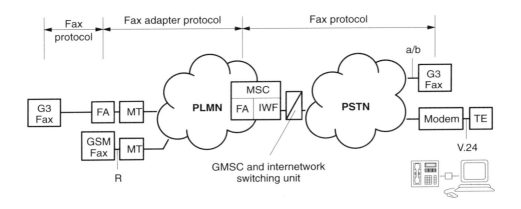

Figure 9.16: Fax adapter in GSM

A more compact version of a mobile fax terminal is possible, if the fax adapter and the Group 3 fax terminal are integrated into a compact terminal (GSM fax, Figure 9.16). Such equipment can be connected at reference point R to an MT2. It delivers a digital signal directly. With this integration, the analog two-wire line interface becomes superfluous. Hence all the analog functions can be omitted such as demodulation and digitalization of the Group 3 fax signals, i.e. the GSM fax needs no analog components such as a modem building block. Therefore, however, the integrated GSM fax must implement the fax adapter protocol and terminate it at reference point R in order to guarantee correct control of the analog fax components in the fax adapter of the IWF.

A complete fax scenario with the required analog components in the fax adapter is illustrated in Figure 9.17. The FA needs several function blocks in the MS as well as in the IWF for the conversion of the fax protocol of the analog a/b interface to the digital transmission procedure in the PLMN. Group 3 fax equipment according to ITU-T standard T.30 employs three modem building blocks, all three operating in half-duplex mode. A V.21 modem (300 bit/s) is used for the signaling phase to set up a fax connection, whereas the information transfer phase uses a V.27ter modem (4.8 kbit/s or 2.4 kbit/s) or a V.29 modem (9.6 kbit/s, 4.8 kbit/s, and 2.4 kbit/s). For conversion from analog signaling tones into messages of the fax adapter protocol, the fax adapter needs an additional *Tone Handler*. It is used to transfer the (re)digitized fax signals over either a transparent or a nontransparent GSM bearer service to the IWF in the MSC. The fax adapter protocol provides a complete mapping of the T.30 protocol, such that the IWF is able to handle the complete fax protocol with the partner entity. From the view of the partner entity, the complete GSM connection consisting of fax adapter, mobile station, and IWF represents a physical connection with a mobile Group 3 fax terminal.

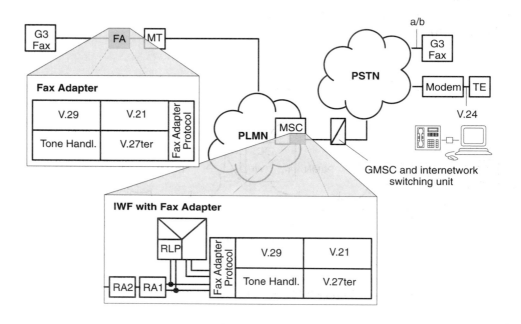

Figure 9.17: Overview of GSM procedure for fax service

Fax service poses special demands on the service quality of the data channel which the GSM network provides for this teleservice. In particular, propagation delays must remain below a maximum threshold, since timers of the T.30 protocol expire otherwise. This is especially critical if RLP is used to reduce transmission errors, which introduces additional delays into the data path through its ARQ procedure.

Two fax services are specified: the transparent procedure and the nontransparent procedure, depending on the kind of bearer service used. The resulting protocol models are shown in Figure 9.18 and Figure 9.19, respectively. It is evident that in both cases the fax protocol is superimposed onto the respective bearer service. The transparent fax procedure is easier to realize, although with its transparent bearer service, it incurs a correspondingly variable fax quality.

In contrast, the nontransparent fax procedure based on RLP is very well protected against transmission errors, and it delivers very acceptable quality of the transmitted documents over a wide range of distances. However, due to the varying transmission conditions, there are also variable delays of the RLP which lead to intermediate buffering of fax signals in the FA and which, in the worst case, can cause the breakdown of the fax transfer [10].

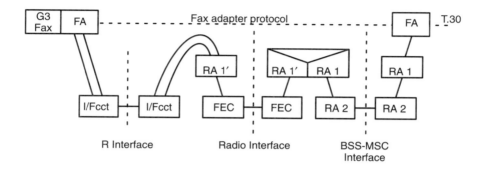

Figure 9.18: Transparent fax procedure in GSM

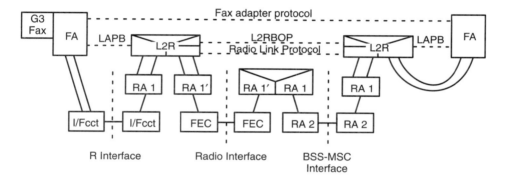

Figure 9.19: Nontransparent fax procedure in GSM

10 Aspects of Network Operation

For the efficient and successful operation of a modern communication network such as a GSM PLMN, a comprehensive *Network Management* (NM) is mandatory. Network management encompasses all functions and activities which control, monitor, and record usage and resource performance of a telecommunication network, with the objective of offering the subscribers telecommunication services of a certain objective level of quality. Various aspects of quality are either defined and prescribed in standards or laid down in operator-specific definitions. Special attention has to be paid to the gap between (mostly simple) measurable technical performance data of the network and the quality of service experienced (subjectively) by the subscriber. Modern network management systems should therefore also include (automated) capabilities to accept reports and complaints from subscribers and convert them into measures to be taken by network management (e.g. trouble ticketing systems).

10.1 Objectives of GSM Network Management

Along with the communication network which realizes the services with its functional units (MS, BSS, MSC, HLR, VLR), one needs to operate a corresponding network management system for support and administration. This NM system is responsible for operation and maintenance of the functional PLMN units and the collection of operational data. The operational data comprise all the measurement data which characterize performance, load, reliability, and usage of the network elements, including times of usage by individual subscribers, which are the basis for calculation of connection fees (billing). Furthermore, in GSM systems in particular, the techniques supporting security must have counterparts in security management functions of network management. This security management is based on two registers: the *Authentication Center* (AUC) provides key management for authentication and encryption and the *Equipment Identity Register* (EIR) provides barring of service access for individual equipment, "blacklisting". To summarize, for all functions of the telecommunication network and its individual functional units (network elements), there are corresponding NM functions.

The GSM standard has defined the following overall objectives of network management:

- International operation of network management
- Cost limitation of GSM systems with regard to short-term as well as long-term aspects
- Achievement of service quality which at least matches the competing analog mobile radio systems

The international operation of a GSM system includes among others the interoperability with other GSM networks (including different countries) and with ISDN networks, as well as the information exchange among network operators (billing, statistical data, subscriber complaints, invalid IMEI etc.). These NM functions are in large part necessary for network operation allowing international roaming of subscribers, and therefore they must be standardized. Their implementation is mandatory.

The costs of a telecommunication system consist of invested capital and operational costs. The investments comprise the cost of the installation of the network and of the network management, as well as development and licensing costs. The periodically incurred costs include operation, maintenance, and administration as well as interest, amortization, and taxes. Lost revenues due to failing equipment or partial or complete network failure must be included in the periodically incurred costs, whereas consequential losses due to cases of failure, e.g. because of lost customers, cannot be estimated and included. Therefore the reliability and maintainability of the network equipment is of course of immense importance and heavily impacts costs. The installation of an NM system on one hand increases the need for investment capital for the infrastructure as well as for spare capacities in the network. On the other hand, these costs for a standardized comprehensive NM system must be compared with the expenses for administration, operation, and maintenance of network elements with manufacturer-proprietary management, or the costs which arise from not recognizing and repairing network failures early enough. Therefore it has to be the objective of a cost-efficient NM system to define and implement uniform vendor-independent network management concepts and protocols for all network elements, and also to guarantee interoperability of network components from different manufacturers through uniform interfaces within the network.

The quality of service to be achieved can be characterized with technical criteria like speech quality, bit error ratio, network capacity, blocking probability, call disconnection rates, supply probability, and availability, and it can also be characterized with nontechnical criteria like ease of operation and comfort of subscriber access or even hot-line and support services.

Considering these objectives, the following functional areas for network management systems can be identified:

- Administrative and business area (subscribers, terminal equipment, charging, billing, statistics)
- Security management
- Operation and performance management
- System version control
- Maintenance

These functions are realized in GSM based on the concept of the *Telecommunication Management Network* (TMN). In general, they are summarized with the acronym FCAPS — *Fault, Configuration, Accounting, Performance, and Security Management* (Figure 10.1).

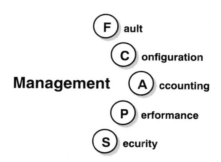

Figure 10.1: Functional areas of TMN systems

Fault management includes functions like failure recognition, failure diagnosis, alarm administration and filtering as well as capabilities for the identification of causes of failures or alarms and keeping of failure logs. Configuration management administers network configurations and handles changes, activates / deactivates equipment, and provides tools for the automatic determination of network topology and connectivity. Accounting management deals with the subscribers and is responsible for the establishment and administration of subscriber accounts and service profiles. Periodic billing for the individual subscribers originates here, based on measured usage times and durations; statistics are calculated, in certain circumstances only for network subareas (billing domains). In performance management, one observes, measures, and monitors performance (throughput, failure rates, response times, etc.), and utilization of network components (hardware and software). The objective is on one hand to ensure a good utilization of resources and on the other hand to recognize trends leading to overload and to be able to start countermeasures early enough. Finally, security management provides for a thorough access control, the authentication of subscribers, and an effective encryption of sensitive data.

10.2 Telecommunication Management Network (TMN)

TMN was standardized within ITU-T / ETSI / CEPT almost simultaneously with the pan-European mobile radio system GSM. The guidelines of the M. series of the ITU-T (M.20, M.30) serve as a framework.

TMN defines an open system with standardized interfaces. This standardization enables a platform-independent multivendor environment for management of all components of a telecommunication network. Essentially it realizes the communication of a management system with network elements it administers, which are considered as managed objects. These objects are abstract information models of the physical resources. A manager can send commands to these administered objects over a standardized interface, can request or change parameters, or be informed by the objects about events that occurred (notification). For this purpose, an agent resides in the managed object, which generates the management messages or evaluates the requests from the manager, and maps them onto corresponding operations or manipulations of the physical resources. This mapping is system specific as well as implementation dependent and hence not standardized. The generalized architecture of a TMN is illustrated in Figure 10.2.

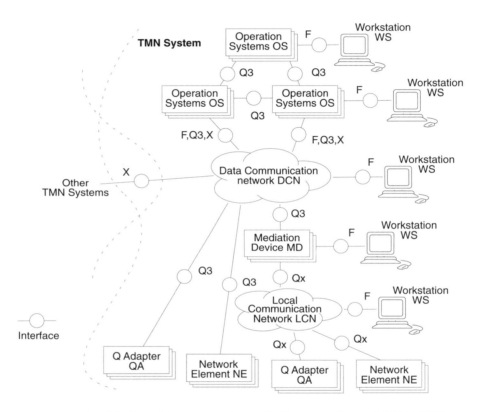

Figure 10.2: TMN architecture (schematically, according to M.3010 [18])

The network management proper is realized in an *Operation System* (OS). The operation systems represent the surveillance and control systems of a TMN system. These systems can communicate with each other directly or form hierarchies. A standardized interface Q3 serves for the communication of the OSs within a TMN, whereas the interconnection of two TMN systems occurs over the X interface (Figure 10.2). The management functionality can also be subdivided into several logical layers according to the OSI hierarchy. For this approach, TMN provides the *Logical Layered Architecture* (LLA) as a framework. The exact numbering and corresponding functionality of each LLA plane were not yet finalized in the standardization process at the time of writing, however, the following planes have been found to be useful (Figure 10.3): *Business Management Layer* (BML), *Service Management Layer* (SML), *Network Management Layer* (NML), and *Element Management Layer* (EML).

The TMN functions of the EML are realized by the network elements NE and contain basic TMN functions such as performance data collection, alarm generation and collection, self diagnosis, address conversion, and protocol conversion. Frequently, the EML is also known as a *Network* EML (NEML) or a *Subnetwork Management Layer* (SNML) [23]. The NML-TMN functions are normally performed by operation systems and used for the realization of network management applications, which require a network-wide scope. For this purpose, the NML receives aggregate data from the EML and generates a global system view from them. On the SML plane, management activities are performed which concern the subscriber and his or her service profile rather than physical network components. The customer contact is administered in the SML, which includes functions like establishing a subscriber account, initializing supplementary services, and several others. The highest degree of abstraction is reached in the BML, which has the responsibility for the total network operation. The BML supports strategic network planning and the cooperation among network operators [23].

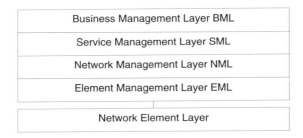

Figure 10.3: Logical layered architecture of a TMN system

For example, an operation system OS could act as a *Basic OS* and be in charge only of a region with a subset of network elements, or it could be a *Network OS* which communicates with all the basic OSs and implements network-wide management functionality. As a *Service OS*, an OS assumes network-wide responsibility for the management of one service, whereas on the BML plane, care is taken of charging, billing, and administration of the whole network and its services.

The individual functional units of the telecommunication network are mapped into *Network Elements* (NEs). These elements are abstract representations of the physical components of the telecommunication network, which is administered by this TMN. The OSs communicate with the network elements over a comprehensive data network, a *Data Communication Network* (DCN). For this purpose, an interface Q3 has been defined, whose protocols comprise all seven layers of the OSI model. However, not every network element must support the full range of Q3 interface capabilities.

For network elements whose TMN interface contains a reduced range of functionality (Qx), a *Mediation Device* (MD) is interposed, which essentially performs the task of protocol conversion between Qx and Q3. A mediator can serve several network elements with incomplete Q3 interfaces, which can be connected to the mediator through a *Local Communication Network* (LCN). The functions of a mediator are difficult to define in general and depend on the respective application, since the range of restrictions of a Qx interface with regard to the Q3 interface is not standardized [12]. Therefore, a mediator could for example realize functions like data storage, filtering, protocol adaptation, or data aggregation and compression.

In spite of ongoing TMN standardization, new network elements and systems without a TMN interface are continuously added and must be integrated. For such cases, the function of the *Q Adapter* (QA) has been defined. In contrast to the mediator MD, which is prefixed to TMN-capable devices with reduced functionality at the Q interface, a QA allows integration of devices which are not TMN capable, and the QA must therefore be tailored for each respective device.

Finally, the operator personnel have access to the TMN system at the F interface through management *Workstations* (WSs) in order to perform management transactions and to check or change parameters. Thus a TMN system gives the network operator at a workstation the capability to supply any network element with configuration data, to receive and analyze failure reports and alarms, or to download locally collected measurement data and usage information. The TMN protocol stack required for this purpose is based on OSI protocols and comprises all seven layers (see also Figure 10.6). The main element of the TMN protocol architecture is the *Common Management Information Service Element* (CMISE) from the OSI system management, which resides in the Application Layer (OSI Layer 7) [23]. The CMISE consists of a service definition, the *Common Management Information Service* (CMIS), and a protocol definition, the *Common Management Information Protocol* (CMIP). The CMISE defines a uniform message format for requests and notifications between management OS and the managed elements NE or the respective QA.

10.3 TMN Realization in GSM Networks

TMN and GSM were standardized approximately at the same time, so that there was a good opportunity to apply TMN principles and methods in a complete TMN system for network management in GSM from the beginning and from ground up. For this purpose, specific working groups were founded for the five TMN categories (Figure 10.1) as well as for architecture and protocol questions which were supposed to develop as much as possible of the TMN system and its services, while following the top-down methodology [18][19] recommended by the ITU-T.

This objective could be pretty much achieved, only that the development methodology was complemented by a bottom-up approach which was rooted in the detailed knowledge about the network components being specified at the same time. The intent was to reach the objective of a complete standard earlier [13][14].

The five TMN categories are essentially realized for all of the GSM system; however, there are some limitations in failure, configuration, and security management. Failure and configuration management are specified only for the BSS; the reasons are that on one hand the databases (HLR, VLR) were assigned to accounting management, and on the other hand standardization efforts were to concentrate on GSM-specific areas. Concentration on GSM-specific areas thereby excluded failure and configuration management for the MSC, which from the management point of view is essentially a standard ISDN switching exchange. For the same reasons, security management is also limited to GSM-specific areas.

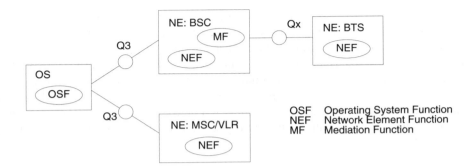

Figure 10.4: A simple TMN architecture of a GSM system (according to [13])

The resulting GSM TMN architecture is shown schematically in Figure 10.4. In GSM, the BSC and the MSC have a Q3 interface as network elements to the OS. Besides the BSS management, the BSC NE always contains a *Mediation Function* (MF) and a Qx interface to the NE supporting the BTS functionality.

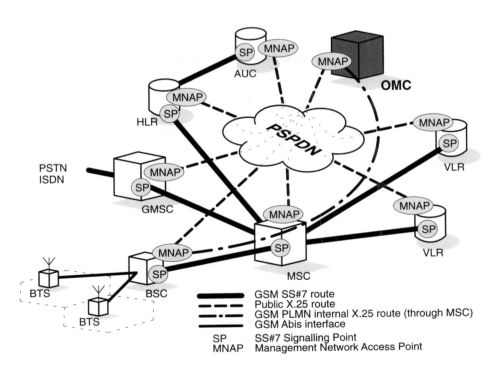

Figure 10.5: Potential signaling interfaces in a GSM TMN

An object-oriented information model of the network has been defined for the realization of the GSM TMN services. The model contains more than 100 *Managed Object Classes* (MOCs) with a total of about 500 attributes. This includes the ITU-T standard objects as well as GSM-specific objects, which include the GSM network elements (BSS, HLR, VLR,

MSC, AUC, EIR) on one hand, but also represent network and management resources (e.g. for SMS service realization or for file transfer between OS and NE) as managed objects. These objects usually contain a state space and attributes which can be checked or changed (request) as well as mechanisms for notification, which report the state or attribute changes. In addition, there are commands for creation or deletion of objects, e.g. in the HLR with *create / modify / delete subscriber* or *create / modify / delete MSISDN* or in the EIR with *create / interrogate /delete equipment* [13]. File transfer objects are used especially in the information model dealing with the registers, since it involves movement of large amounts of data.

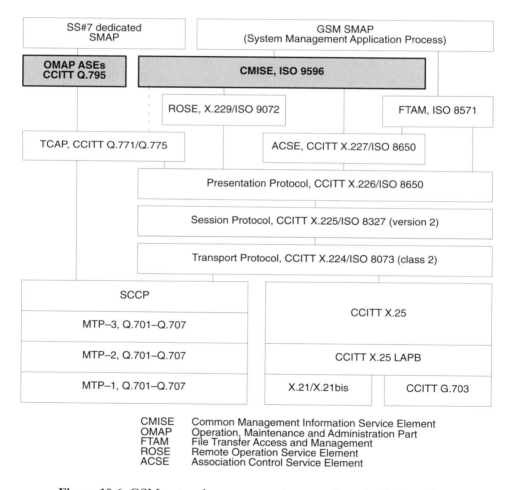

Figure 10.6: GSM network management protocols at the Q3-interface

The TMN communication platform to be used as *Data Communication Network* DCN can be either an OSI X.25 packet network or the SS#7 signaling network (MTP and SCCP). Both offer a packet switching service which can be used to transport management messages. Each network element is connected to this management network over a *Management Network Access Point* (MNAP); see Figure 10.5.

If the TMN uses X.25, the DCN can be the public PSPDN or a dedicated packet switching network within the PLMN with the MSC as a packet switching node. In addition, the MSC can include an interworking function for protocol conversion from an external X.25 link to the SS#7 SCCP, which realizes the connection of the OMC to the PLMN through an external X.25 link. Further transport of management messages is then performed by the SS#7 network internal to the PLMN.

The framework defined for the GSM TMN protocol stack at the Q3 interface is presented in Figure 10.6. The end-to-end transport of messages between OS and NE is realized with the OSI Class 2 transport protocol (TP2), which allows the setup and multiplexing of end-to-end transport connections over an X.25 or SCCP connection. Error detection and data security are not provided in TP2; they are not needed since X.25 as well as SCCP offer a secure message transport service already.

Of course, the OSI protocol stack also needs the protocols for the data link and presentation layers. The OSI *Common Management Information Service Element* (CMISE) plays the central role in GSM network management. Its services are used by a *System Management Application Process* (SMAP) to issue commands, to receive notifications, to check parameters, etc. For file transfer between objects, GSM TMN uses the OSI *File Transfer Access and Management Protocol* (FTAM). It is designed for the efficient transport of large volumes of data.

CMISE needs a few more *Service Elements* (SEs) in the application layer for providing services: the *Association Control Service Element* (ACSE) and the *Remote Operations Service Element* (ROSE). The ACSE is a sublayer of the application layer which allows application elements (here CMISE) to set up and take down connections between each other. The ROSE services are realized with a protocol which enables initiation or execution of operations on remote systems. This way ROSE implements the paradigm also known as *Remote Procedure Call* (RPC).

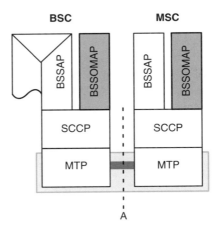

Figure 10.7: Operation and maintenance of the BSS

There is also a management system for the signaling components of a GSM system. This SS#7 SMAP uses the services of the *Operation Maintenance and Administration Part* (OMAP) which allows observation, configuration, and control of the SS#7 network resources. Essentially, the OMAP consists of two *Application Service Elements* (ASEs), the *MTP Routing Verification Test* (MRVT) and the *SCCP Routing Verification Test* (SRVT) which allow verification of whether the SS#7 network works properly on the MTP or SCCP planes. Another *Management Application Part* is the *Base Station Operation and Maintenance Application Part* (BSSOMAP) which is used to transport management messages from OMC to BSC through the MSC over the A interface and to execute management activities for the BSS (Figure 10.7, and compare it with Figure 7.11) [24].

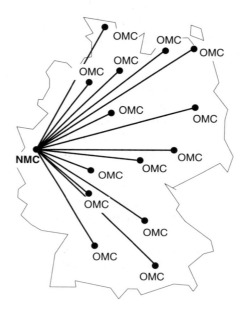

Figure 10.8: Hierarchical organization of network management within Germany

Network management is usually organized in a geographically centralized way. For the remote surveillance and control of network management functions there are usually one or more *Operation and Maintenance Centers* (OMCs). For efficient network management, these OMCs can be operated as regional subcenters according to the LLA hierarchy of the various TMN management planes, and they can be combined under a central *Network Management Center* (NMC); see Figure 10.88.

11 GSM – What Next?

11.1 Globalization

Six years after the start of operation of the first commercial GSM networks, the proportion of GSM of worldwide mobile communication networks had grown to about 28% and still shows increasing tendency. This increasing globalization has lead to several new measures. The PCS / PCN networks are using new frequencies around 1800 MHz, and in North America around 1900 MHz. International roaming among these networks is possible based on the standardized interface between mobile equipment and the SIM, which enables personalization of equipment operating in different frequency ranges (SIM card roaming). Furthermore, a more general standardization of the SIM concept could allow worldwide roaming across non-GSM networks.

Besides roaming based on the SIM card, the MoU places increasing emphasis on multiband systems and multiband terminals. Multiband systems permit the simultaneous operation of base stations with different frequency ranges. In connection with multiband terminals, this approach leads to a powerful concept which is clearly pushed by the MoU. Such terminals can be operated in several frequency bands, and they can adapt automatically to the frequencies used in the network at hand. This enables roaming among networks with different frequency ranges, but also automatic cell selection in multiband networks with different frequencies becomes possible.

11.2 GSM Services in Phase 2+

GSM is not a closed system that does not undergo change. The GSM standards continue to be developed further; in the current phase of standardization (Phase 2+) more than 80 individual topics are being discussed [20]. Phase 1 of the GSM implementation contained basic teleservices − in the first place voice communication − and a few supplementary services, which had to be offered by all network operators in 1991 when GSM was introduced into the market. The standardization of Phase 2 was completed in 1995 with market introduction following in 1996. Essentially, ETSI added more of the supplementary services, which had been planned already when GSM was initially conceived and which were adopted from the fixed ISDN net (see Section 4.3). These new services made it necessary to rework large portions of the GSM standards. For this reason, networks operating according to the revised standard are also called GSM Phase 2 [20]. However, all networks and terminals of Phase 2 preserve the compatibility with the old terminals and network equipment of Phase 1, i.e. all new standard development had to be strictly backward compatible.

The subjects considered deal with many aspects ranging from radio transmission to communication and call processing. However, there is no complete revision of the GSM standard; rather single subject areas are treated as separate standardization units, with the intent of allowing them to be implemented and introduced independently from each other. Thus GSM systems can evolve gradually, and in the future, standardization can meet market needs in a flexible way. However, with this approach a unique identification of a GSM standard version becomes impossible. The designation GSM Phase 2+ is supposed to indicate this openness [20], suggesting an evolutionary process with no endpoint in time or prescribed target dates for the introduction of new services.

A large menu of technical questions is being addressed of which only a few are presented as examples in the following. It should be emphasized that these services were only in the planning or discussion stage at the time of writing; actual implementations were not yet certain.

11.2.1 Teleservices

GSM Phase 2 defined essentially a set of new supplementary services. Now Phase 2+ is addressing new bearer and teleservices, too. One of the most important services in GSM is (of course) voice service. Thus it is obvious, that voice service has to be further developed. In first place is the development of new speech codecs with two competing objectives: better utilization of the frequency bands assigned to GSM as well as improvement of speech quality in the direction of the quality offered by ISDN networks, which is primarily requested by professional users.

The reason for improved bandwidth utilization is to increase the network capacity and the spectral efficiency, i.e. traffic carried per cell area and frequency band. For voice services, early plans were already in place to introduce a half-rate speech codec at a later time. Under normal conditions, this device achieves, in spite of half the bit rate, almost the same speech quality as the full-rate codec used so far. However, quality loss occurs for mobile-to-mobile communication (tandem mobile calls), because in this case due to the ISDN architecture one has to go twice through the GSM speech coding/decoding process. The end-to-end through-transmission of GSM-coded speech is intended to avoid multiple unnecessary transcoding and the resulting quality loss (Figure 11.1) [20].

The other important concern is the improvement of speech quality. Speech quality that is close to the one found in fixed networks is especially important for business applications and in cases where GSM systems are intended to replace fixed networks, e.g. for fast installation of telecommunication networks in areas with insufficient or missing telephone infrastructure. Work on the *Enhanced Full Rate codec* (EFR) is therefore considered of high priority. This EFR is a full-rate codec; Nevertheless, it achieves speech quality clearly superior to the current full-rate codec. In the North American DCS1900 networks, such an EFR is already standardized and in use. This EFR is considered a good candidate for standardization by ETSI [20].

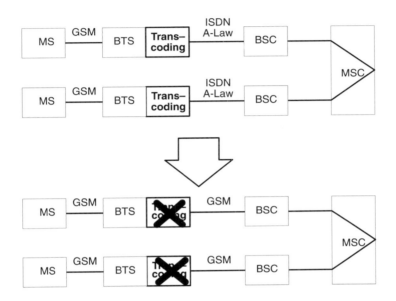

Figure 11.1: Through-Transport of GSM-coded speech in Phase 2+

The groupcall is another teleservice currently under discussion. This service feature has been known from mobile radio systems, and is strongly requested by railroad operators. In 1992 their international organization, the *Union Internationale des Chemins de Fer* (UIC), selected the GSM system as their standard [20]. The groupcall is offered in its own service area, which can comprise several cells. To realize the service, traffic channels are not allocated only to individuals but used by a group of users together. In each case only one group member can transmit (speaker) whereas all the others receive (Figure 11.2). In each participating cell, only one full-rate channel is occupied (as in regular voice calls), i.e. the voice signal of the speaker is broadcast to all listening participants in a cell on one group channel. The right to talk can be passed along within the group during a group call by using a push-to-talk mechanism as in mobile radio, which is supported by a special signaling facility.

In addition, development continues with data services. Two prominent trends can be recognized: integration of packet services into GSM networks and high bit rate bearer services with data transmission rates up to 100 kbit/s [20].

Packet-switched data transmission is already possible and standardized in existing GSM networks. It consists of offering access to the *Public Switched Packet Data Network* (PSPDN); see Sections 9.5.3 and 9.6.2. However, on the air interface such access to a PSPDN occupies a complete circuit-switched traffic channel. For bursty traffic this leads to inefficient resource utilization. The newly defined packet data service, *General Packet*

Radio Service (GPRS), offers a genuine packet-switched bearer service at the air interface. Two variations are offered, similar to ISDN [22] packet data communication. The first version uses a Bm channel (GPRS proper) and the second version goes on the Dm channel (packet data on signaling). The first version aims for a realization where a traffic channel is shared in competition by several GPRS participants. The main problem in this mode is to offer multiple access with fair sharing and avoidance of collisions. The second version can be considered as a further simplified evolution of the *Short Message Service* (SMS). In this case the already existing transport procedures for SMS can also be used for packet-switched user data across the air interface.

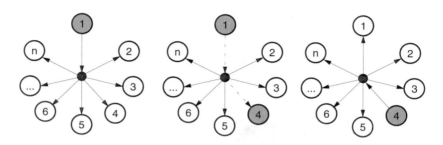

Figure 11.2: Groupcall scenarios

The currently offered maximal data rate of 9600 bit/s for data services is rather low compared to fixed networks. The desire for higher data rates in GSM networks is therefore quite obvious. Accordingly, one of the GSM standardization groups is investigating a *High-Speed Circuit-Switched Data* (HSCD) service. By combining several traffic channels, one expects to achieve data rates of up to 100 kbit/s. Whereas this can be relatively easily achieved at the base stations, the changes required on the terminal side are substantial. An HSCD-capable terminal must not only be able to transmit and receive simultaneously on several time slots, it must also supply the considerable signal processing power for modulation / demodulation and channel coding. To use this new HSCD service thus requires a new generation of mobile stations with significantly increased capabilities.

11.2.2 Supplementary Services

A lot of the standardization efforts in GSM Phase 2+ are concerned with expanding supplementary services. By far the largest part of the supplementary service characteristics known from ISDN have in one way or another already been implemented in GSM (see Section 4.3). The mobility of the users necessitates new supplementary services, but mobility frequently also makes their implementation much more difficult. Examples of new supplementary services being now standardized for GSM and known from ISDN or newly defined are *Mobile Access Hunting, Short Message Forwarding, Multiple Subscriber Profile, Call Transfer*, or *Completion of Calls to Busy Subscribers* (CCBS).

The example of the CCBS service shows especially clearly how much the role of the HLR is changing from its original function as a database to a more active role as a service control component, similar to the *Service Control Point* (SCP) of the *Intelligent Network* (IN). The supplementary service CCBS basically realizes "Call Back if Busy". If a called subscriber momentarily does not accept a call due to an ongoing conversation, the calling subscriber can activate the supplementary service CCBS which causes the network to notify him at the end of the called subscriber's ongoing call and automatically set up the new connection. The subscriber mobility adds more complexity to the implementation of this service. In the fixed network, implementation would require the establishment of queues for call-back requests in the switching center of the calling and called subscribers, respectively. In a mobile network, this may involve additional switches, because after activation of the CCBS service, the calling subscriber may be roaming into another switching center area. If the implementation of the service were only in the MSC, either there could be a central-ized solution or the queuing lists would have to be forwarded to the new MSC − which may even be in another network. The targeted solution is centralized in the HLR, which has to store, in addition to the current MSC designation, the callback queues, if any, of the subscriber. If the mobile station changes the MSC area, the callback queue is trans-ferred to the new MSC. In this case, therefore, the HLR has to assume an additional server role and perform call control beyond the originally planned restriction to a pure database function.

11.3 GSM and Intelligent Networks

The procedures for the development of the GSM standards required close cooperation of the involved manufacturers and network operators. The international standardization of services and interfaces led to a set of common successful performance characteristics in GSM networks, most prominently the international roaming capability. In contrast, the network operators desire service differentiation to be able to gain competitive advan-tages. The more a performance criterion is standardized, the lower are the costs of devel-opment and introduction on one hand, but on the other hand the standardization of ser-vices and service performance criteria also reduces the possibility for differentiation among competitors. Moreover, the time-to-market is often pushed out because of the pro-longed process of standardization.

The current approach to GSM standardization attempts to overcome this dilemma. Instead of specifying services and supplementary services directly or completely, only mechanisms are standardized which enable introduction of new services. With this ap-proach it should be possible to restrict the implementation of a service to a few switches in the home network of a subscriber, whereas local (visited) switches have to provide only a fixed set of basic functions and the capability to communicate with the home network switch containing service logic. This group of GSM standards within Phase 2+ is known under the name *Support of Operator-Specific Services* (SOSS), or also as *Customized Ap-plications for Mobile Network Enhanced Logic* (CAMEL). Essentially, this effort repre-sents a convergence of GSM and *Intelligent Network* (IN) technologies.

The fundamental concept of IN is to enable flexible implementation, introduction and control of services in public networks and to use the idea of dividing the switching func-tionality into basic switching functionality, residing in *Service Switching Points* (SSPs) and centralized service control functionality, residing in *Service Control Points* (SCPs). Both network components communicate with each other over the signaling network using the generic SS#7 protocol extension called *Intelligent Network Application Part* (INAP). This approach enables a centralized, flexible, and rapid introduction of new services [33].

There are already some features in GSM which parallel an Intelligent Network. Even though GSM standards use neither IN terminology nor IN protocols, i.e. INAP, the GSM network structure follows the IN philosophy [21]. In the GSM architecture, the separation into functional units like MSC and HLR and the consistent use of SS#7 and its MAP extensions are in conformity with the IN architecture's split into SSPs and SCPs which communicate using INAP.

The philosophy of CAMEL is to proceed with the implementation of services in GSM in a similar way as in IN. This is reflected in the separation into a set of basic call processing functions in the MSC or GMSC which act as SSP on one hand, and on the other hand the intelligent service control functions in the home network of the respective subscriber. The HLR in a GSM network already has functions similar to the SCP, especially with regard to supplementary services. Beyond that, the CAMEL approach provides its own dedicated SCPs. Imagine specialized SCPs for the translation of abbreviated numbers in *Virtual Private Networks* (VPN) or for future extended *Short Message Services* (SMSs). With this configuration, the service implementation with its service logic is needed only once, namely in the home network SCP. The network operator offering the service thus has the sole control over the features and performance range of the service. Because of the complete range of generic functions that have to be provided at each SSP (MSC, GMSC, etc.), new services can immediately be provided in each network, and an uninterrupted service availability is guaranteed for roaming subscribers. The sole responsibility and control for the introduction of new services lies in SOSS / CAMEL with the operator of the home network, the contract partner of the subscriber. This opens new competitive possibilities among network operators. Operator-specific services can be introduced rapidly without having to go through the standardization process, and yet they are available worldwide.

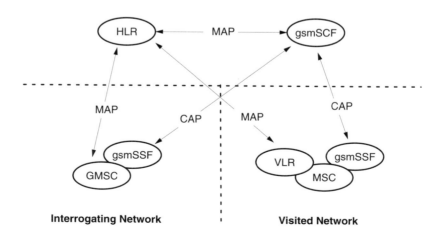

Figure 11.3: Functional architecture for CAMEL

Figure 11.3 shows the resulting architecture. The CAMEL specification requires a GSM-specific version of IN. Similar to the IN approach, GSM defines a basic call processing function as a *GSM Service Switching Function* (gsmSSF) and a service logic function *GSM Service Control Function* (gsmSCF). In addition to the MAP signaling interfaces already existing in Phase 1 and Phase 2 for communication between visited and home networks (GMSC, VLR, HLR), new signaling interfaces are needed for communication between basic switching and service logic in the visited and home networks. For this signaling, a new Application Part of SS#7 is being specified, the *CAMEL Application Part* (CAP), which thus assumes the function similar to INAP. These functions and protocols represent the basic structure for the realization of intelligent services and their flexible introduction.

The prerequisites for CAMEL are the definition of a standardized extended call model with appropriate trigger points, and the specification of the generic range of services which must be provided by the SSP. More precisely, the new extended call model must also include a model of subscriber behavior, because besides normal call processing aspects, it also contains events like *Location Updating*. For each subscriber, this behavior model is stored in the HLR and supplied from the home network to the currently visited SSP / MSC. In this way, the SSP / MSC has a set of trigger points with corresponding SCP addresses for each subscriber roaming in its area. When a trigger condition is satisfied, the call and transaction processing in the SSP / MSC is interrupted, and the SCP is notified. The SCP can now analyze the context and give instructions to perform certain functions to the SSP according to the service implementation. Typically these might be call forwarding, call termination, or other stimuli to the subscriber [20]. Based on this behavior model and the corresponding control protocol between mobile SSP and home SCP, which are connected through a set of generic SSP functions, we can expect in the future to see a large variety of operator-specific services.

11.4 Beyond GSM: On the Road to UMTS

With all its enhancements, GSM will long represent the mainstream of mobile communication systems. However, it is obvious, that because of technical and economical reasons, GSM will be followed by a third generation of mobile communication systems. This system, dubbed *Universal Telecommunication System* (UMTS), within ETSI/Europe and *International Mobile Telecommunication 2000* (IMT2000) within ITU, is aimed to support a wide range of voice and nonvoice services, narrowband and wideband, focussing on data communication services (packet-switched data) based on IP technology. UMTS shall give the mobile user similar performance as in the fixed network and will stimulate the development of new multimedia applications.

Looking at the rapidly growing GSM subscriber numbers, it can be also foreseen that any future system must support a very high number of subscribers. Penetration rates of up to 50% are to be expected. With respect to the radio spectrum needed for the evolving mass market and considering the bandwidth requirements of the envisaged broadband services (up to 2 Mbit/s), the radio interface has to be more spectrum-efficient than today. Therefore European countries and others have devoted considerable effort to developing the concepts for a flexible and efficient next generation of mobile communication systems.

In contrast to earlier evolution scenarios, and taking into account the worldwide success of GSM, UMTS will build, as much as possible, on the existing GSM infrastructure and techniques. However, it will have a new radio interface, the *UMTS Terrestrial Radio Access* (UTRA), using the frequency bands around 2000 MHz and new multiple access techniques. In January 1998, ETSI made the basic decision on the radio interface. Two interfaces were proposed [38]

- In the so-called paired band, using *Frequency Division Duplex* (FDD), the system adopts the radio access technique Wideband-CDMA (W-CDMA).

- In the so-called unpaired band, using *Time Division Duplex* (TDD), the UMTS system adopts the radio access technique called TD-CDMA, essentially a combination of TDMA and CDMA. A detailed description can be found in [39].

The UTRA proposal has worldwide support. Among its backers are the following companies. From the equipment manufacturer's side by Alcatel, Bosch, Ericsson, Italtel, Motorola, NEC, Nokia, Nortel, Siemens, Sony. From the operators side by Deutsche Telekom, France Telecom, Mannesmann Mobilfunk, NTT DoCoMo, Telecom Finland, Telia, and Vodafone.

According to ETSI, "the agreed solution offers a competitive continuation for GSM to UMTS." The decision in the ITU on IMT2000 is expected to be made in 1999. Equally important as the radio interface will be the service concept of UMTS. With respect to the service aspects, the standard will provide [38] two sorts of mechanisms:

- Mechanisms to enable the creation of supplementary services including the creation and execution of appropriate MMI (man–machine interface) procedures to the user's terminal.

- Mechanisms to permit the definition of interworking functions, appropriate for the creation of teleservices and/or end-user applications, including the downloading and execution of these functions in the user's terminal and in appropriate network elements.

Overall, it is the intention of all participants in the UMTS standardization process to enable a smooth transition from second-generation GSM to third-generation UMTS/IMT2000 systems.

References

[1]   Kleinrock, L. *Queueing Systems — Vol. 1 Theory*. New York: J. Wiley, 1975

[2]   Tran-Gia, P. *Analytische Leistungsbewertung verteilter Systeme*. Berlin: Springer, 1996

[3]   Eberspächer, J. (ed.): *Sichere Daten, sichere Kommunikation — Secure Information, Secure Communication*. Publication of Münchner Kreis, Telecommunications, vol. 18. Berlin: Springer, 1994

[4]   Vögel, H.-J., Johr, H. and Grom, A. *Messung und verbesserte Markov-Modellierung transparenter GSM-Datendienste*. In: B. Walke (ed.) *Mobile Kommunikation*, Lectures of the ITG-Fachtagung, Sept. 1995, Neu-Ulm. Berlin: vde-Verlag, 1995 (ITG Technical Report 135) pp. 279–287

[5]   Mende, W. Bewertung ausgewählter Leistungsmerkmale von zellularen Mobilfunksystemen. Dissertation, Hagen Fernuniversität, 1991

[6]   Junius, M. and Marger, X. Simulation of the GSM handover and power control based on propagation measurements in the German D1 network. In: *Proceedings of the fifth Nordic Seminar on Digital Mobile Radio Communications (DMR V)*, Helsinki, 1992, pp. 367–372

[7]   Fuhrmann, W., Brass, V., Janßen, U., Kühl, F. and Roth, W. Digitale Mobilkommunikationsnetze. In Gerner, N., Hegering, H.-G. and Swoboda, J. (eds.) *Kommunikation in verteilten Systemen*, Tutorium anläßlich der ITG/GI-Fachtagung Kommunikation in verteilten Systemen KiVS, March 1993, Munich. Lehrstuhl f. Datenverarbeitung, Technische Universität München, 1993

[8]   Xu, G. and Li, S.-Q. Throughput multiplication of wireless LANs for multimedia services: SDMA protocol design. In *Proceedings of Globecom 94*, Nov./Dec. 1994, San Francisco. New York: IEEE, 1994, pp. 1326–1332.

[9]   Gottschalk, H. Zeichengabetechnische Anbindung digitaler Mobilfunknetze an das Festnetz der Telekom. In: Walke, B. (ed.) *Informationstechnische Gesellschaft im VDE: Mobile Kommunikation*, Lectures of the ITG-Fachtagung, Sept. 1993, Neu-Ulm. Berlin: vde-verlag, 1993 (ITG Technical Report 124)

[10]   Decker, P. and Pertz, U. Simulative Leistungsbewertung der nichttransparenten Fax-Übertragung im GSM-System. In: Walke, B. (ed.) *Informationstechnische Gesellschaft im VDE: Mobile Kommunikation*, Lectures of the ITG-Fachtagung, Sept. 1993, Neu-Ulm. Berlin: vde-verlag, 1993 (ITG Technical Report 124)

[11]   Natvig, E. Evaluation of six medium bitrate coders for the pan-European digital mobile radio system. *IEEE Journal on Selected Areas in Communications*, vol. 6, no. 2, 1988, pp. 324–334

[12]   Glitho, R. H. and Hayes, S. Telecommunications management network: vision
        vs. reality. *IEEE Communications Magazine*, vol. 33, no. 3, 1995, pp. 47–52

[13]   Towle, T. S. TMN as applied to the GSM Network. *IEEE Communications Mag-
        azine*, vol. 33, no. 3, 1995, pp. 68–73

[14]   Lin, Yi-Bing OA&M for the GSM Network. *IEEE Network Magazine*, vol. 11,
        no. 2, 1997, pp. 46–51

[15]   ISO/IEC 33091991. Information Technology – Telecommunications and infor-
        mation exchange between systems – High-level data link control (HDLC) pro-
        cedures – Frame structure

[16]   ITU-T Recommendation E.164. Numbering Plan for the ISDN Era

[17]   ITU-T Recommendation V.110. Support of Data Terminal Equipment (DTEs)
        with V-Series Type Interfaces by an Integrated Services Digital Network (ISDN)

[18]   CCITT Recommendation M.3010. Principles for a Telecommunications Ma-
        nagement Network, Genf, 1992

[19]   CCITT Recommendation M.3020. TMN Interface Specification Methodology,
        Genf, 1992

[20]   Mouly, M. and Pautet, M.-B. Current evolution of the GSM systems. *IEEE Per-
        sonal Communications Magazine*, Oct. 1995, pp. 9–19

[21]   Laitinen, M. and Rantale, J. Integration of intelligent network services into fu-
        ture GSM networks. *IEEE Communications Magazine*, June 1995, pp. 76–86

[22]   Bocker, P. *ISDN- das diensteintegrierende digitale Nachrichtennetz*. 3d ed., Berlin:
        Springer, 1990

[23]   Sahin, V. Telecommunications management metwork – principles, models and
        applications. In: Aidarous, S. and Plevyak, T. (eds.) *Telecommunications Net-
        work Management into the 21st Century*. New York: IEEE Press, 1993, pp. 72–121

[24]   Schmidt, S. Management (Operation & Maintenance) von GSM Base Station
        Subsystemen. In: Walke, B. (ed.) *Informationstechnische Gesellschaft im VDE:
        Mobile Kommunikation*, Lectures of the ITG-Fachtagung, Sept. 1993, Neu-Ulm.
        Berlin: vde-verlag, 1993 (ITG Technical Report 124)

[25]   Lee, W. C. Y. *Mobile Cellular Telecommunication Systems*. New York: McGraw-
        Hill, 1989

[26]   Proakis, J. G. *Digital Communications*. 2d ed., New York: McGraw-Hill, 1989

[27]   Bertsekas, D. and Gallager, R. *Data Networks*. Englewood Cliffs, NJ: Prentice
        Hall, 1987

[28]   Bossert, M. D-Netz-Grundlagen – Funkübertragung in GSM-Systemen, Teil 1
        und 2. *Funkschau*, vol. 22 and 23, 1991

[29]   Steele, R. *Mobile Radio Communications*. London: Pentech Press, 1992

[30]   Watson, C. Radio Equipment for GSM. In: Balston, D.M. and Macario, R.C.V
        (eds.). *Cellular Radio Systems*. Norwood, MA: Artech House, 1993

[31]   Bossert, M. *Kanalcodierung*. Stuttgart: B.G. Teubner Verlag, 2d ed., 1998

[32]   David, K. and Benkner, T. *Digitale Mobilfunksysteme*. Stuttgart: B.G. Teubner
        Verlag, 1996

[33]   Ambrosch, Maher and Sasseer *The Intelligent Network*. Berlin: Springer, 1989

[34]   Hagenauer, J. and Seshadri, N. The performance of rate compatible punctured convolutional codes. *IEEE Transactions on Communications*, vol. 38, 1990, pp. 966–980

[35]   Begin, G. and Haccoun, D. Performance of sequential decoding of high-rate punctured convolutional codes. *IEEE Transactions on Communications*, vol. 42, no. 3, 1994, pp. 966–987

[36]   Kallel, S. Complementary punctured convolutional (CPC) codes and their applications. *IEEE Transactions on Communications*, vol. 43, no. 6, 1995, pp. 2005–2009

[37]   Hartmann, Ch. and Vögel, H.-J. Teletraffic analysis of SDMA-systems with inhomogeneous MS location distribution and mobility. *Wireless Personal Communications*, Special Issue on Space Division Multiple Access. Dordrecht: Kluwer Academic Press, 1998.

[38]   http//www.etsi.org

[39]   *IEEE Communications Magazine*, Special Issue on ACTS Mobile Program in Europe. vol. 36, no. 2, 1998, pp. 80–136

Appendix A: Selected GSM standards

Here is a selection of the most important GSM standards (GSM recommendations).

[i]   GSM-Rec. 01.06, Service Implementation Phases

[ii]   GSM-Rec. 02.16, International MS Equipment Identities

[iii]   GSM-Rec. 02.17, Subscriber Identity Modules − Functional Characteristics

[iv]   GSM-Rec. 02.78, Digital Cellular Telecommunications System (Phase 2+); Customized Applications for Mobile Network Enhanced Logic (CAMEL); Service Definition (Stage 1)

[v]   GSM-Rec. 03.02, Network Architecture

[vi]   GSM-Rec. 03.03, Numbering, Addressing and Identification

[vii]   GSM-Rec. 03.04, Signaling Requirements Relating to Routing of Calls to Mobile Subscribers

[viii]   GSM-Rec. 03.08, Organization of Subscriber Data

[ix]   GSM-Rec. 03.09, Handover Procedures

[x]   GSM-Rec. 03.10, GSM PLMN Connection Types

[xi]   GSM-Rec. 03.12, Location Registration Procedures

[xii]   GSM-Rec. 03.20, Security Related Network Functions

[xiii]   GSM-Rec. 03.45, Technical Realization of Facsimile Group 3 Service (Transparent)

[xiv]   GSM-Rec. 03.46, Technical Realization of Facsimile Group 3 Service (Non transparent)

[xv]   GSM-Rec. 03.78, Digital Cellular Telecommunications System (Phase 2+); Customized Applications for Mobile Network Enhanced Logic (CAMEL) − Stage 2

[xvi]   GSM-Rec. 04.02, GSM PLMN Access Reference Configuration

[xvii]   GSM-Rec. 04.04, MS-BSS Layer 1 General Requirements

[xviii]   GSM-Rec. 04.05, MS-BSS Data Link Layer − General Aspects

[xix]   GSM-Rec. 04.06, MS-BSS Data Link Layer Specification

[xx]   GSM-Rec. 04.08, Mobile Radio Interface Layer 3 Specification

[xxi]   GSM-Rec. 04.10, Mobile Radio Interface Layer 3 Supplementary Services
           Specification − General Aspects

[xxii]   GSM-Rec. 04.21, Rate Adaptation on the MS-BSS Interface

[xxiii]   GSM-Rec. 04.22, Radio Link Protocol for Data and Telematic Services on the
           MS-BSS Interface

[xxiv]   GSM-Rec. 04.80, Mobile Radio Interface Layer 3 Supplementary Services
           Specification − Formats and Coding

[xxv]   GSM-Rec. 05.01, Physical Layer on the Radio Path (General Description)

[xxvi]   GSM-Rec. 05.02, Physical Layer on the Radio Path

[xxvii]   GSM-Rec. 05.03, Channel Coding

[xxviii]   GSM-Rec. 05.04, Modulation

[xxix]   GSM-Rec. 05.05, Radio Transmission and Reception

[xxx]   GSM-Rec. 05.08, Radio Sub-System Link Control

[xxxi]   GSM-Rec. 05.10, Radio Sub-System Synchronization

[xxxii]   GSM-Rec. 06.01, Speech Processing Functions: General Description

[xxxiii]   GSM-Rec. 06.10, GSM Full-Rate Speech Transcoding

[xxxiv]   GSM-Rec. 06.11, Substitution and Muting of Lost Frames for Full-Rate Speech
           Traffic Channels

[xxxv]   GSM-Rec. 06.12, Comfort Noise Aspects for Full-Rate Speech Traffic Channels

[xxxvi]   GSM-Rec. 06.31, Discontinuous Transmission (DTX) for Full-Rate Speech
           Traffic Channels

[xxxvii]   GSM-Rec. 06.32, Voice Activity Detection

[xxxviii] GSM-Rec. 07.07, AT Command Set for GSM Mobile Equipment

[xxxix]   GSM-Rec. 08.02, BSS/MSC Interface Principles

[xl]   GSM-Rec. 08.08, BSS-MSC: Layer 3 Specification

[xli]   GSM-Rec. 08.09, Network Management Signaling Support Related to the BSS

[xlii]   GSM-Rec. 08.20, Rate Adaptation on the BSS-MSC Interface

[xliii]   GSM-Rec. 08.52 BSC to BTS Interface Principles

[xliv]   GSM-Rec. 08.58, BSC-BTS: Layer 3 Specification

[xlv]   GSM-Rec. 09.01, General Network Interworking Scenarios

[xlvi]   GSM-Rec. 09.02, Mobile Application Part

[xlvii]   GSM-Rec. 09.03, Requirements on Interworking between the ISDN or PSTN
           and the PLMN

[xlviii] GSM-Rec. 09.05, Interworking between the PLMN and the PSPDN for PAD Access

[xlix]   GSM-Rec. 09.07, General Requirements on Interworking between the PLMN and the ISDN or PSTN

[l]   GSM-Rec. 09.78, Digital Cellular Telecommunications System (Phase 2+); CAMEL Application Part (CAP) specification

[li]   GSM-Rec. 11.11, Specification of the SIM-ME Interfaces

[lii]   GSM-Rec. 12.00, Objectives and Structures of Network Management

[liii]   GSM-Rec. 12.01, Common Aspects of GSM Network Management

[liv]   GSM-Rec. 12.02, Subscriber, Mobile Equipment and Service Data Administration

[lv]   GSM-Rec. 12.03, Security Management

[lvi]   GSM-Rec. 12.04, Performance Data Measurement

[lvii]   GSM-Rec. 12.05, Subscriber Related Event and Call Data

[lviii]   GSM-Rec. 12.06, GSM Network Change Control

[lix]   GSM-Rec. 12.07, Operations and Performance Management

[lx]   GSM-Rec. 12.10, Maintenance Provisions for Operational Integrity of Mobile Stations

[lxi]   GSM-Rec. 12.11, Maintenance of the Base Station System

[lxii]   GSM-Rec. 12.13, Maintenance of the Mobile Services Switching Centre

[lxiii]   GSM-Rec. 12.14, Maintenance of Location Registers

[lxiv]   GSM-Rec. 12.20, Network Management Procedures and Messages

[lxv]   GSM-Rec. 12.21, Network Management Procedures and Messages on the Abis Interface

Appendix B: Addresses in GSM

BCC	Base Station Colour Code
BSIC	Base Station Identity Code
(M)CC	(Mobile) Country Code
CI	Cell Identifier
FAC	Final Assembly Code
GCI	Global Cell Identity
IMSI	International Mobile Subscriber Identity
LAC	Location Area Code
LAI	Location Area ID
LMSI	Local Mobile Subscriber Identity
MNC	Mobile Network Code
MSIN	Mobile Subscriber Identification Number
MSISDN	Mobile Station ISDN Number
MSRN	Mobile Station Roaming Number
NCC	Network Colour Code
NDC	National Destination Code
NMSI	National Mobile Subscriber Identity
SN	Subscriber Number
SNR	Serial Number
TAC	Type Approval Code
TMSI	Temporary Mobile Subscriber Identity

	TAC	FAC	SNR	SP	(M)CC	MNC	MSIN	SN	LAC	CI	NCC	BCC
IMEI	▓	▓	▓	▓								
IMSI					▓	▓	▓					
MSISDN					▓			▓				
MSRN					▓			▓				
LAI					▓	▓			▓			
CI										▓		
BSIC											▓	▓

Appendix C: Acronyms

3PTY	Three Party Service
A3, A5, A8	Encryption Algorithms
AB	Access Burst
Abis	BTS - BSC Interface
ACCH	Associated Control Channel
ACSE	Association Control Service Element
ADPCM	Adaptive Delta Pulse Code Modulation
AGCH	Access Grant Channel
AOC	Advice of Charge
ARQ	Automatic Repeat Request
ASE	Application Service Element
ATM	Asynchronous Transfer Mode
AUC	Authentication Center
BA	BCCH Allocation
BAIC	Barring of All Incoming Calls
BAOC	Barring of All Outgoing Calls
BC	Bearer Capability
BCC	Base Station Colour Code
BCCH	Broadcast Control Channel
BCF	Base Control Function
BCH	Broadcast Channel
BFI	Bad Frame Indication
BIC-Roam	Barring of Incoming Calls when Roaming Outside the Home PLMN
Bm	Mobile B-Channel
BML	Business Management Layer
BN	Bit Number
BOIC	Barring of Outgoing International Calls
BOIC-exHC	Barring of Outgoing International Calls except those to Home PLMN
BSC	Base Station Controller
BSIC	Base Station Identity Code
BSS	Base Station Subsystem
BSSAP	Base Station System Application Part
BSSMAP	Base Station System Management Application Part

BSSOMAP	Base Station Operation and Maintenance Application Part
BTS	Base Transceiving Station
BTSM	Base Transceiving Station Management
C/R	Command/Response
CA	Cell Allocation
CAMEL	Customized Applications for Mobile Network Enhanced Logic
CAP	CAMEL Application Part
CBCH	Cell Broadcast Channel
CC	Country Code
CCBS	Completion of Call to Busy Subscriber
CCCH	Common Control Channel
CDMA	Code Division Multiple Access
CEP	Connection End Point
CELP	Code Excited Linear Predictive Speech Coding
CEPI	Connection End Point Identifier
CEPT	Conference Européenne des Adminstrations des Postes et des Télécommunications
CFB	Call Forwarding on Mobile Subscriber Busy
CFNRc	Call Forwarding on Mobile Subscriber Not Reachable
CFNRy	Call Forwarding on No Reply
CFU	Call Forwarding Unconditional
CI	Cell Identifier
CLIP	Calling Number Identification Presentation
CLIR	Calling Number Identification Restriction
CM	Connection Management
CMISE	Common Management Information Service Element
CODEC	Coder/Decoder
COLP	Connected Line Identification Presentation
COLR	Connected Line Identification Restriction
CONF	Conference Calling
CRC	Cyclic Redundancy Check
CT	Call Transfer
CUG	Closed User Group
CW	Call Waiting
DB	Dummy Burst
DCCH	Dedicated Control Channel
DCN	Data Communication Network
DECT	Digital Enhanced Cordless Telecommunication
Dm	Mobile D-Channel
DLCI	Data Link Connection Identifiers
DRX	Discontinuous Reception
DTAP	Direct Transfer Application Part
DTMF	Dual-Tone Multifrequency

DTX	Discontinuous Transmission
EFR	Enhanced Full Rate (codec)
EIR	Equipment Identity Register
EML	Element Management Layer
ETSI	European Telecommunication Standards Institute
FA	Fax Adapter
FAC	Final Assembly Code
FACCH	Fast Associated Control Channel
FB	Frequency Correction Burst
FCAPS	Fault, Configuration, Accounting, Performance, Security
FCCH	Frequency Correction Channel
FDD	Frequency Division Duplex
FDMA	Frequency Division Multiple Access
FEC	Forward Error Correction
FH-CDMA	Frequency Hopping CDMA
FN	TDMA Frame Number
FPH	Freephone Service
FPLMTS	Future Public Land Mobile Telecommunication System
FTAM	File Transfer Access and Management
GCI	Global Cell Identity
GMSC	Gateway MSC
GMSK	Gauss Minimum Shift Keying
GPRS	General Packet Radio Service
GSC	GSM Speech Codec
GSM	Global System for Mobile Communication
gsmSCF	GSM Service Control Function
gsmSSF	GSM Service Switching Function
HDLC	High-Level Data Link Control
HLR	Home Location Register
HSCD	High-Speed Circuit Switched Data
HSN	Hopping Sequence Number
I/Fcct	Interface Circuit
IE	Information Elements
IFP	Interface Protocol
IMEI	International Mobile Equipment Identity
IMSI	International Mobile Subscriber Identity
IN	Intelligent Network
INAP	Intelligent Network Application Part
ISC	International Switching Center
IWF	Interworking Function
Kc	Cipher/Decipher Key
Ki	Subscriber Authentication Key
L2ML	Layer 2 Management Link

L2R	Layer 2 Relay
L2RBOP	Layer 2 Relay Bit-Oriented Protocol
L2RCOP	Layer 2 Relay Character-Oriented Protocol
LA	Location Area
LAC	Location Area Code
LAI	Location Area ID
LAN	Local Area Networks
LAPB	Link Access Procedure B
LAPD	Link Access Procedure D
LAPDm	Link Access Procedure D mobile
LAR	Log Area Ratios
LCN	Local Communication Network
LEO	Low Earth Orbiting satellites
LI	Length Indicator
LLA	Logical Layered Architecture
LMSI	Local Mobile Subscriber Identity
LPC	Linear Predictive Coding
LTP	Long-Term Prediction
MA	Mobile Allocation
MAH	Mobile Access Hunting
MAIO	Mobile Allocation Index Offset
MAP	Mobile Application Part
MC	Multi-Carrier
MCC	Mobile Country Code
MCI	Malicious Call Identification
MD	Mediation Device
MEO	Medium Earth Orbiting satellites
MHS	Message Handling System
MM	Mobility Management
MMI	Man-Machine Interface
MNC	Mobile Network Code
MOC	Managed Object Classes
MOS	Mean Opinion Score
MoU	Memorandum of Understanding
MRVT	MTP Routing Verification Test
MS	Mobile Station
MSC	Mobile Switching Center
MSIN	Mobile Subscriber Identification Number
MSISDN	Mobile Station ISDN Number
MSRN	Mobile Station Roaming Number
MT	Message Type, Mobile Termination
NB	Normal Burst
NCC	Network Colour Code

NDC	National Destination Code
NE	Network Element
NM	Network Management
NMC	Network Mangement Center
NML	Network Management Layer
NMSI	National Mobile Subscriber Identity
NMT	Nordic Mobile Telephone
NTP	Nontransparent Protocol
NUI	Network User Identification
OACSU	Off-Air Call Setup
OMAP	Operation, Maintenance and Administration Part
OMC	Operation and Maintenance Center
OML	Operation and Maintenance Link
OMSS	Operation and Maintenance Subsystem
OS	Operation System
P-IWMSC	Packet Interworking MSC
PAD	Packet Assembler/Disassembler
PCH	Paging Channel
PCM	Pulse Code Modulation
PCN	Personal Communication Network
PCS	Personal Communication System
PDN	Public Data Network
PDU	Protocol Data Unit
PLMN	Public Land Mobile Network
PN	Pseudorandom Number
PSPDN	Public Switched Packet Data Network
QA	Q-Adapter
QN	Quarter Bit Number
RA	Random Access Procedure
RA	Rate Adaptation
RACH	Random Access Channel
RAND	Random Number
REVC	Reverse Charging
RFCH	Radio Frequency Channel
RFN	Reduced TDMA Frame Number
RLP	Radio Link Protocol
ROSE	Remote Operation Service Element
RPE	Regular Pulse Excitation
RPE-LPT	Regular Pulse Excitation − Long-Term Prediction
RR	Radio Resource Management
RSL	Radio Signaling Link
RTLL	Radio in the Local Loop
SACCH	Slow Associated Control Channel

SAP	Service Access Point
SAPI	Service Access Point Identifier
SB	Synchronization Burst
SC	Service Center
SCCP	Signaling Connection Control Part
SCH	Synchronization Channel
SCN	Sub-Channel Number
SCP	Service Control Point
SDCCH	Standalone Dedicated Control Channel
SDMA	Space Division Multiple Access
SDU	Service Data Unit
SE	Service Element
SID	Silence Descriptor
SIM	Subscriber Identity Module
SM-CL	Short Message Control Layer
SM-CP	Short Message Control Protocol
SM-RL	Short Message Relay Layer
SM-RP	Short Message Relay Protocol
SM-TL	Short Message Transport Layer
SM-TP	Short Message Transport Protocol
SMAP	System Management Application Process
SMC	Short Message Control
SMG	Special Mobile Group
SML	Service Management Layer
SMNL	Subnetwork Management Layer
SMR	Short Message Relay
SMS	Short Message Service
SMS-GMSC	Short Message Service – Gateway MSC
SMS-IWMSC	Short Message Service – Interworking MSC
SMS-SC	Short Message Service – Service Centre
SMSCB	Short Message Service Cell Broadcast
SMSS	Switching and Management Subsystem
SN	Subscriber Number
SNR	Serial Number
SNR	Signal-to-Noise Ratio
SOSS	Support of Operator-Specific Services
SP	Spare
SP	Signaling Point
SPC	Signaling Point Code
SRES	Signature Response
SRVT	SCCP Routing Verification Test
SS	Supplementary Services
SSP	Service Switching Point

STC	Sub Technical Committees
TA	Terminal Adapter, Timing Advance
TAC	Type Approval Code
TACS	Total Access System
TCAP	Transaction Capabilities Application Part
TCH	Traffic Channel
TDD	Time Division Duplex
TDMA	Time Division Multiple Access
TE	Terminal Equipment
TEI	Terminal Equipment Identifier
TETRA	Trans European Trunked Radio
TMN	Telecommunication Management Network
TMSI	Temporary Mobile Subscriber Identity
TN	Time Slot Number, TDMA Frame Number
TRAU	Transcoding and Rate Adaptation Unit
TSC	Training Sequence Code
TXF	Transceiver Function
UDI	Unrestricted Digital Information
Um	Air Interface
UMTS	Universal Mobile Telecommunication System
UNI	User−Network Interface
UPT	Universal Personal Telecommunication
UUS	User-to-User Signalling
VAD	Voice Activity Detection
VLR	Visited Location Register, VLR Number
WLL	Wireless Local Loop
WS	Workstation

Index

B